NATIONAL AGRICULTURAL BIOTECHNOLOGY COUNCIL REPORTS

NATIONAL AGRICULTURAL BIOTECHNOLOGY COUNCIL

1991

[redacted]

NABC REPORT 3

Agricultural Biotechnology at the Crossroads

Biological Social & Institutional Concerns

EDITED BY JUNE FESSENDEN MACDONALD

NATIONAL AGRICULTURAL BIOTECHNOLOGY COUNCIL
ITHACA, NEW YORK

NATIONAL AGRICULTURAL BIOTECHNOLOGY COUNCIL

The National Agricultural Biotechnology Council provides an open forum for the discussion of issues related to the impact of biotechnology on agriculture. The views presented and positions taken by individual participants in this report are their own and do not necessarily reflect the views or policies of the NABC or The Joyce Foundation, sponsors of this report.

ADDITIONAL COPIES ARE AVAILABLE FOR $7. PLEASE MAKE CHECKS OR PURCHASE ORDERS PAYABLE TO:

NABC / BTI
159 Biotechnology Building
Cornell University
Ithaca, NY 14853-2703

National Agricultural Biotechnology Council, Ithaca, NY 14853-2703

ISBN: 0-9630907-4-0 700 paper

Printed on recycled paper by Union Press of Binghamton, New York.

The National Agricultural Biotechnology Council is a consortium of not-for-profit agricultural research and educational institutions established in 1988.

Providing an open forum for exploring issues in agricultural biotechnology.

NABC'S PRINCIPAL OBJECTIVES ARE TO:

- *provide an open forum for persons with different interests and concerns to come together to speak, to listen, to learn and to participate in meaningful dialog and evaluation of the potential impacts of agricultural biotechnology*
- *define issues and public policy options related to biotechnology in the food, agricultural and environmental areas*
- *promote increased understanding of the scientific, economic, legislative and social issues associated with agricultural biotechnology by compiling and disseminating information to interested people*
- *facilitate active communication among researchers, administrators, policy makers, practitioners and other concerned people to insure that all viewpoints contribute to the safe and efficacious development of biotechnology for the benefit of society.*
- *sponsor meetings and workshops and publish and distribute reports that provide a foundation for addressing issues*

ACKNOWLEDGEMENTS

NABC gratefully acknowledges the funding for this volume and the Third Annual NABC Meeting from The Joyce Foundation.

Special recognition goes to the University of California Agricultural Issues Center, especially Lynne Kennedy and John Woolcott, the meeting coordinators and Sandy Fisher, who handled registration plus many other of the details. L.C. "Bees" Butler and the planning committee, Hal Carter, Jim Grieshop, Martin Kenney, Martina McGloughlin, Jim Murray, Carolyn Napoli, and Cal Qualset at the University of California–Davis, the host institution, deserve special thanks as do workshop facilitators Jim Brenner, Rick Bennett, Jim Reedy and Elli Rilla from the North Bay Public Policy and Education Team.

Special thanks goes to Hana Barker, Design/Editorial Coordinator, NABC, who not only provided the overall design for this report, but also careful organizational oversight for the meeting. Thanks also to Kate O'Hara who stepped in as NABC Design/Editorial Coordinator to complete the production of this report, including both suggesting and drawing the illustrations.

The aerial photographs were provided courtesy of the Cornell Laboratory for Environmental Applications of Remote Sensing (CLEARS), Center for the Environment.

PREFACE

In the four years since Ralph W. F. Hardy, President of the Boyce Thompson Institute for Plant Research and Robert B. Nicholas Esq., McDermott, Will & Emery, developed the concept of a consortium of not-for-profit agricultural research and educational institutions concerned about ensuring the safe development of agricultural biotechnology for the benefit of all sectors of society, the National Agricultural Biotechnology Council (NABC) has grown into a truly national organization.

JUNE FESSENDEN MACDONALD
Editor

A member of the faculty in the Section of Biochemistry, Molecular and Cell Biology, and the Biology and Society Program at Cornell University, she is also Deputy Director of NABC.

The Third Annual NABC Meeting established NABC as filling an important and heretofore unmet need in agricultural biotechnology by providing an open forum for people holding diverse views to meet, talk, listen and learn from one another. The meeting (NABC 3), organized in cooperation with the University of California Agricultural Issues Center, provided an opportunity for various stakeholders to take a look back over the last decade and assess what had and had not happened in agricultural biotechnology and why, to review current directions and to look forward at the paths to be chosen. Participants were asked to 1–assess the constraints and incentives that impact research, commercialization and acceptance of biotechnology products, 2–identify the factors that facilitate or impede the application of biotechnology in agriculture and 3–seek ways to assess their appropriateness.

It is hoped that this report, which communicates the results of that meeting—the lively, even contentious discussions in the workshops with their conclusions and recommendations and the diverse viewpoints (13 plenary lectures and 4 background talks and papers) which provided background material for those workshops—will provide new information to some, a foundation for addressing the many concerns raised for others and an increased understanding of divergent viewpoints for all.

The first section provides an overview of "agricultural biotechnology at the crossroads," highlighting the issues identified by the various speakers and the workshop recommendations on which there was consensus among the participants. The editor takes responsibility for the summarization of these issues and recommendations with the hope that no one view or perspective was slighted. The second section presents the full reports with all of the recommendations developed in each "national issues" workshop. The following sections provide more in-depth and specific information on visions of the future; biological breakthroughs and bottlenecks; institutional incentives and impediments and socioeconomic issues and perspectives as well as the background papers provided to workshop participants in advance of the meeting.

NABC hopes that this volume, with its wide range of perspectives on four timely topics—herbicide tolerance in crops, biological control of pests, animal growth promotants and transgenic animals— will provide the basis for further dialog and consensus building on the future directions and role of biotechnology in U.S. agriculture.

The photo essays presented in this volume, on the front and back covers as well as divisions between the sections, were selected as another illustration of still different crossroads that agriculture has faced and is facing from other technologies. These aerial photographs taken at between 11 and 16 year intervals are of two different farmland areas being impacted by development. It is against this landscape that the new road of molecular biotechnology is being constructed.

CONTENTS

AGRICULTURAL BIOTECHNOLOGY AT THE CROSSROADS

C.1938

AGRICULTURAL BIOTECHNOLOGY AT THE CROSSROADS *AN OVERVIEW*

The National Agricultural Biotechnology Council's Third Annual Meeting was held in cooperation with the University of California Agricultural Issues Center at the end of May, 1991. Focusing on the theme "Agricultural Biotechnology at the Crossroads," the meeting offered the opportunity to assess the current status and potential future impacts of agricultural biotechnology. The last decade saw many people project agricultural biotechnology's potential to raise productivity—yet that reality has not yet been achieved. Why?

More importantly, what does and should the future hold for creating an environment in which agricultural biotechnology can safely and acceptably be used to enhance productivity in agriculture?

The Third Annual Meeting was organized to address these questions and to:

—promote dialog and understanding among those with differing views;

—establish a common knowledge base and identify areas of disagreement both of fact and perception;

—reach consensus where possible on specific areas needing additional research, education or study;

—develop criteria for evaluating the appropriateness of constraints and incentives currently impeding or promoting the research, commercialization and acceptance of products;

—develop specific recommendations, including policy options.

NABC/JOYCE FELLOWS

ROLEY PIGGOTT

Agricultural Economics
UC Davis — Piggott is studying the cross-market impacts of biotechnology.

JEFFREY C. FEARN

Plant Production
Boyce Thompson Inst.
Fearn is working to characterize sym5 pea nodulation mutant.

NABC/JOYCE GRADUATE FELLOWS

CASSANDRA KLOTZ

Agricultural Economics
UC Davis — Klotz is researching the adoption and diffusion of bovine somatotropin in the California dairy industry.

JOHAN F. SWINNEN

Agricultural Economics
Cornell University
Swinnen is studying the decision-making process in agricultural policy setting by different governments.

The meeting clearly promoted an exchange of views among individuals from diverse backgrounds, including animal, plant and social scientists, administrators, various interest groups (farmers, consumers, environmentalists and animal welfare groups) and entrepreneurs. With each NABC meeting the number and diversity of participants expands.

An even mix of lectures and workshop sessions provided participants with many opportunities to speak, listen and learn. The lectures included visions of agricultural biotechnology's future and assessments of the status quo of various biotechnologies, institutional concerns and socioeconomic issues. Presentations and white papers on four "national issues" provided a common foundation for the four workshops—*herbicide tolerance in crops, biological control of pests, transgenic animals* and *animal growth promotants.*

MAJOR CURRENTS

Over the course of two and half days, several themes emerged:

Agricultural biotechnology should not be viewed as an end in itself; rather, as a set of tools which might be used to achieve societal goals, whatever those might be. Margaret Mellon's talk, "Biotechnology and the Environmental Vision" (see page 66) precipitated discussions on this area throughout the meeting. The choice and development of agricultural production systems should be made with society's environmental goals in mind.

Keynote speaker Walter Truett Anderson's talk "The Past and Future of Agriculture," (see page 53) focused on the increasing awareness of *how human actions affect the environment.* Concern for the environment and limited global resources, along with an increasing ability to measure and detect these changes, will play a major role in determining acceptable agricultural production decisions in the future, necessitating long–term planning. Many at the meeting felt that *agricultural research funding should be strengthened.* Secure funding compatible with the nature of the research (i.e., commitment of funds for 10-year periods rather than the usual one to three year period) may assist in shifting the national planning horizon to the long term. At the same time, socioeconomic studies designed to reveal societal preference patterns, among other things, also should be funded.

Preserving the "family farm" was often mentioned, particularly in the workshops, as a desirable social goal which is believed to be at risk as agricultural biotechnologies become more widespread. However, the family farm is threatened by other sources more imminent than those posed by agricultural biotechnology—for instance, the high cost of current farm programs.

In order to *remain competitive in world markets*, the United States may need to have a mixture of agricultural technologies available. Developments in other parts of the world may hold serious implications for U.S. agricultural trade, particularly those involving the use of agricultural biotechnologies.

The *regulatory systems involved with agricultural biotechnology need to be clearly delineated*, available to the public and industry and subject to continual scrutiny. The regulatory system should not stifle basic research and should be protective of human and animal welfare.

VISIONARY PATHWAYS

The keynote speaker, journalist, political scientist and author Walter Truett Anderson suggested that agricultural biotechnology will have an impact comparable to the industrial revolution on today's productivity and life-styles. Urging those with vested interests to engage in open discussions on the potential impact of agricultural biotechnology, he emphasized the need to abandon "satanic" views of biotechnology in favor of examining what uses, both economic and humanitarian, we have for these tools. He reminded us that the history of humankind has continually been one of adapting the environment to suit particular purposes and that it is appropriate to consider the emergence of agricultural biotechnology as but one event in a chronology of discoveries which commenced when humans first started planting grains.

Reiterating stages proposed by NABC Chair Ralph W. F. Hardy in his charge to the meeting, Anderson spoke of our passing through a stage of "active science" (1860s – early 1970s) and a "transitional era" (early 1970s – early 1990s) and now of entering a "product era," characterized by large risks and large rewards, where products based on agricultural biotechnology will be developed. How well has society coped with the evolutionary process? Anderson says it is "a mixed bag," with some logical responses as well as some rather chaotic ones.

Margaret G. Mellon and Jerry Caulder followed Anderson with alternative views on the future of agricultural biotechnology. For Mellon, the Director of the National Biotechnology Policy Center, National Wildlife Federation, agricultural biotechnology should be examined within the context of broader social goals—fundamental goals such as chemical–free agriculture should be high on the agenda for public debate. She presented a vision of prosperous farmers, an abundance of safe and affordable food and a clean environment, suggesting that biotechnology was dampening our progress towards a sustainable agriculture. Mellon suggested that the agri–biotechnology industry was indeed facing a crossroads regarding commercialization, adding that credible regulatory programs are needed now if transgenic products are to reach the marketplace.

Caulder, President and CEO of Mycogen Corporation and Chair of the Industrial Biotechnology Association, enthusiastically spoke of the future of agricultural biotechnology, noting we are on the verge of having an impressive portfolio of products. He cited increased production efficiency as a path that will benefit all consumers, adding that the public needs to be better informed about the associated risks and benefits with new products. Caulder believes that agriculture would be more at risk if biotechnology is abandoned than if biotechnology becomes commonplace. Noting that the 60 day food "surplus" in the world is really an inadequate "reserve," he called for more funding in the agricultural biotechnology area for basic research to insure that the best scientists are attracted to the area.

ON THE FRONTIERS

Looking at the frontiers of biotechnology, ten speakers attempted to answer the questions, "Where are we now? What factors are, can and should impede progress?"

Covering biological breakthroughs and bottlenecks, Calvin Qualset, the Director of the Genetic Resources Conservation Program at the University of California–Davis, Winston J. Brill, principal of Winston J. Brill & Associates Consulting and George E. Seidel Jr., an animal physiologist at Colorado State University, all spoke about the "cutting edge" of biotechnology and its potential applications, reminding us of the frustrations and obstacles (biological, financial and regulatory) facing rapid development and release of new agricultural biotechnologies. Time delays associated with field testing are but one example.

The next four speakers broadened the scope of discussion, lending their expertise to examine issues beyond the laboratory—the institutional incentives and impediments to agricultural biotechnology. Robert B. Nicholas, a partner at the Washington law firm of McDermott, Will & Emery and Executive Director, NABC, called for a shorter and less costly regulatory framework based on the products of agricultural biotechnology rather than on the research–discovery process. Guidelines should be clear regarding jurisdiction in the various areas, a system which the public trusts, and should allow for and encourage discussion.

Offering an assessment of the incentives related to property rights, William H. Lesser, an agricultural economist at Cornell University, pointed out that companies have little incentive to invest in research or product development without sufficient reassurance that they can recoup research and development costs through ownership of resulting processes and products. On the other hand, university scientists who conduct much of the research have strong incentives to publish research results and university administrators often want ownership of those discoveries which might have commercial possibilities. Lesser identified several areas where intellectual property laws should be reexamined to "achieve a better balance between private incentive and public well-being". Examples include the laws regarding the patenting of "pioneering discoveries" and the U.S. "first-to-invent" system. However, he felt that the state of intellectual property protection in the United States is quite extensive and its shortcomings are not principally responsible for the slow commercialization of products.

Roger Salquist, Chair and CEO of Calgene, Inc., followed with a discussion of some of the problems associated with commercialization, identifying the slow nature of the science as a serious "structural impediment". Lack of funding, from both the private and public sectors, also slows research and development as well as the "convoluted" structure of agriculture resulting from government interventions—although he did not feel that the criticism of regulatory agencies regarding procedures and delays for obtaining field trial approvals is warranted.

The public sector's role in agricultural biotechnology was addressed by Charles Hess, the Assistant Secretary for Science and Education at the United States Department of Agriculture (USDA), noting that as agribusiness becomes more market-oriented, it is the university and government laboratories that must play an ever increasing role in doing research in biotechnology that may not lead to a definitive product. A strong national commitment to funding for agricultural biotechnology is essential for U.S. competitiveness. It is also in the public's interest to insure that the best and the brightest women and men are attracted to and remain in agricultural science. Taking exception with those who maintain biotechnology would compete with sustainable agriculture, Hess maintained that "biotechnology can help agriculture be sustainable, productive and nutritious."

Economic issues also factor into decisions to develop and use agricultural biotechnologies. Susan Offutt, Chief, Agriculture Branch at the Office of Management and Budget, presented two situations which could have a profound influence on agricultural production decisions. One, farming will become more competitive if the current round of General Agreement on Tariffs and Trades (GATT) negotiations results in a dismantling of domestic farm programs and freer international trade in agricultural products. She anticipates that farmers will respond by diversifying output mixes more than at present. Two, the Clean Water Act is due for re–authorization during the current session of Congress, and attention will focus on farming's contribution to surface water quality degradation. The "price" of using the natural environment will likely rise. In both cases, the choice of inputs and outputs will be weighed more carefully by farmers. The products of biotechnology must show a positive contribution to agriculture if biotechnology is to be accepted.

William Lacy, the Assistant Dean for Research, Pennsylvania State University, presented the fourth criterion as an essential element in evaluating future research and developments in biotechnology—and asked if it should really be considered the "first criterion." He emphasized that biotechnology presents tools to move toward social and economic goals, but also has the potential to increase inequities among groups in the U.S. and between developed and developing countries. Lacy suggested we therefore need the fourth criterion as part of our decision-making process.

The final speaker, Michael Hansen, Project Leader at the Consumers Union, believes that consumers are better informed about developments in agricultural biotechnology than is commonly thought. Noting that the consumer will and should be demanding information on products from agricultural biotechnology, he said that consumers rank drug companies as the least credible source of information about food quality and safety. Critical of several public agency procedures, Hansen called for more governmental and industrial accountability to the public.

THE WORKSHOPS: NATIONAL ISSUES

Four workshops offered all participants the opportunity to speak, to listen, and to learn. White papers (see page 177) were distributed to workshop registrants prior to the meeting as background material for each workshop and general information on each topic was presented during the meeting in a session on "national issues." A feature common to all workshops was the variety of backgrounds and interests of the participants.

Herbicide tolerance in crops—this workshop addressed the questions, "What is the probable role of herbicide tolerant crops in agricultural production?" and "What are the benefits and risks associated with their use?"

Biological control of pests—this workshop developed strategies for a national effort to make biological control viable for farmers, consumers and agribusiness, as well as safe for the environment.

Transgenic animals—a less developed area, this workshop examined the technical difficulties, achievements and physiological consequences of producing transgenic animals. Associated moral and ethical issues, and health and food safety concerns were raised.

Animal growth promotants—this workshop had the goal of assessing the biological, socioeconomic, health and safety, environmental quality and communications problems associated with the commercialization of animal growth promotants.

Recommendations developed in these workshops can be found on pages 11 - 15 and full summaries on pages 27 - 49.

CROSSROADS

While workshops and lectures helped to identify areas of common ground between participants, there were also many areas where consensus was not developed, most noticeably in the area of specific solutions. Many questions were left unanswered. "What is an appropriate level of research funding? Who should set and monitor ethical standards in relation to research? Who should carry out impact studies? Who should be regulated? How? Who should address international concerns, and how?"

Although many questions remain, this meeting provided the unique opportunity for people from various constituencies with different values and agendas to begin to forge new relationships with groups previously considered opponents, and to begin to understand some of the common, and not-so-common, concerns.

Patricia B. Swan, Interim Provost, Iowa State University, summed up the meeting, identified both roads which have already been crossed, and many which are yet to come. Roads crossed include the definition of agricultural biotechnology as "a set of methods or tools used in living systems" as well as the acceptance that these tools will be used and regulated (see page 16).

Crossroads facing us now include "What will drive the future of agriculture and agricultural biotechnology" and "Will technology, economics or social concerns dominate widespread regulatory concerns; insuring communication among the public, the scientists, the agribusiness sector and the regulators; and the issue of funding." One of the most problematic crossroads identified was "Who gets to decide these questions?" which only serves to emphasize the importance of open forums such as NABC3.

In closing, Swan asked, "How vigorous and fearless was our examination of the issues?," adding that while she had seen some "fairly courageous sifting and winnowing in the last two days," the discussions developed at the meeting should continue. There is a need for more cooperative exploration of the roads ahead.

Agricultural Biotechnology at the Crossroads was a crossing of many visions and values in agricultural biotechnology. With this brief meeting, many points of view have come to the fore, hopefully initiating discussions and understandings and the next steps toward consensus building on the roles of biotechnology in agriculture.

WORKSHOP RECOMMENDATIONS

HERBICIDE TOLERANCE IN CROPS

Concentrating on four priority issues and goals, this workshop made the following recommendations:

HTCs and herbicide use—There was general agreement that the primary goal of research should be to ensure that HTCs result in the safer use of safer herbicides. Additionally, farmers must learn how to integrate the various options for weed control into the best management plan for their farms.

Conduct field research studies that compare alternative weed control strategies, including how HTCs will affect the overall production system.

Focus HTC research on herbicides with the following attributes: low toxicity to non–target species, including humans and wildlife; low residues in the environment, including groundwater, surface water and air; non toxic residues in crops and food; low use rates; appropriate degradation to benign breakdown products; cost effectiveness; compatibility with alternative weed management strategies; compatibility with technology improvements in the way herbicides are applied; and increased reliability of weed control accompanied by improved crop yields.

Health and environmental risk assessment—There was agreement that any risks need to be identified, and criteria to assess these risks developed. The federal government should:

Evaluate environmental data for HTCs on a case–by–case basis and in a reasonable time frame.

Use public funds to construct a data base from information obtained from small scale field trials.

Provide guidance to ensure that seed companies and other institutions developing HTCs ensure the food safety and quality of the crops.

Socioeconomic Impacts—*Government institutions that currently deal with socioeconomic concerns should:*

Consider the impact of HTCs and, if appropriate, mitigate any impacts.

Foster public discussions on the impacts of HTCs, including socioeconomic impacts.

It was agreed that socioeconomic impacts of HTCs should not be a "fourth criterion" for regulatory approval of a product for commercialization—the products should be judged on three criteria: human and environmental safety, quality and efficacy.

Regulatory Policy—Participants felt that the impact of regulations on the cost of products and the development of minor versus major crops must be considered. A fair, timely process clarifying or establishing regulatory policy is needed. The federal government should:

Articulate a policy stating what is regulated and who regulates by January 1, 1992. Within a reasonable time after a regulatory policy has been articulated, it should be "tested" with a specific HTC ready to go through such a process. Three general criteria for this process were set: 1- provide meaningful opportunities for public input; 2- allow industry to proceed in a fair and timely manner; and 3- as appropriate, responsible health and environmental safety reviews should be conducted.

BIOLOGICAL CONTROL OF PESTS

Out of four subgroups identified, (technology, regulation, commercialization and adoption), four issues emerged as important for the development and increased application of biological control in the coming years.

First, because of a *lack of leadership,* biological control must get on the national agenda. For this to happen, advocacy is necessary from academicians, legislators and other public policy makers.

Second, *effective and reasonable regulatory procedures* are needed.

Third, a *tax on pesticides* should be established to provide government revenues for the promotion of biological control applications.

Fourth, an *increase in government funding* in the public arena (research and extension) and the *establishment of (financial) incentives* for an increase in supply (industry) and demand (farmers) of biological control is necessary.

Each subgroup in the workshop developed specific recommendations:

Technology—Establish and maintain basic and applied research programs to address scientific issues. In order of priority, research programs should focus on: 1- host-pest biocontrol agent interactions; 2-ecological relationships between the target pest, its environment and biological control agent; 3-host resistance mechanisms; and 4-compatibility of biological control agents with chemical agents in integrated pest management programs.

Provide public funding and incentives for 'public good' (commercial products) types of biological control.

Provide incentives for 'private good' types of biological control. Among the variety of options discussed with no consensus: diverting money from the Clean Water Act; a pesticide tax; an "Orphan Drug Act" for small market biopesticides; R&D tax credits; lowering capital gains taxes to help R&D investments; and regulations for field testing and registration of biocontrol agents that are clear and concise.

Regulation—In order of priority:

Redefine regulatory procedures by defining agency responsibility for organism groups; define criteria/characteristics representing risks and benefits, and establishing a fixed time for regulatory decisions to on-field test applications.

Improve communication by facilitating access to federal and state regulatory procedures, by establishing voluntary mechanisms to share the results of safety tests between investigators and between federal and state authorities (such as the National Biological Impact Assessment Program (NBIAP)data base in the USDA).

Commercialization—Incentives to increase use of biologics and minimize the current barriers are necessary:

Set up national research centers to develop biological control methods with locals and cooperatives, and create a clearinghouse for basic and applied information on the delivery of biological control agents.

Modify crop support programs to encourage diversification and provide insurance premiums against crop loss to farmers who use biological control agents.

Develop programs to change the lack of tolerance for imperfections at all levels of the marketing system.

Adoption and Implementation—Establish a national program promoting biological control which includes developing educational and informational materials and their distribution, and establish demonstration projects on farms.

Establish programs to reduce the initial cost disadvantage of biological control agents over traditional pest management techniques, possibly including taxing pesticides/users and use of the revenues for these programs.

TRANSGENIC ANIMALS

Four general areas were identified for discussion:

Unintended and unwanted consequences—Listed were: 1- the impact on genetic diversity, animals as disease vectors and effects upon wild populations as well as pollution that may be associated with animal production; 2- the well-being of the animals themselves, both in terms of their health and their ability to lead relatively tranquil lives; and 3- consequences not readily anticipated.

Conduct animal genetic research to enhance human and animal well-being at an acceptable risk to animals, humans and the environment. Strategies expected to achieve this goal: 1- sufficient public funds to conduct research in the area of transgenics, thereby reducing the pressure to rush to market and to support research on risk; 2- the use of peer review panels, such as current animal care committees, to assure proper procedures are used by researchers and the public is made aware of these procedures and has the opportunity to express concern to the appropriate bodies; and 3- the development of model systems to anticipate both unwanted and beneficial outcomes of transgenic research.

Socioeconomic concerns—In order to *minimize the negative and maximize the positive socioeconomic impacts of transgenic animals and their products at the local, national and international levels,* four key steps were identified: 1- determine the type of transgenic animal or product, the time frame in which it will be developed and adopted, and where and how it will be used; 2- identify direct economic consequences (good and bad) of adoption, including downstream second and third-order consequences; 3- identify possible associated environmental benefits/costs; and 4- identify segments of world society that are affected by transgenic animal technology.

Funding—Since transgenic technology is merely a tool, it should not be singled out as a special priority for funding. Rather, *basic animal genetic research funding should be increased in order to increase the genetic knowledge base.*

Food safety and consumer acceptance—Participants felt that the existing research and regulatory systems need to be more effective in building public communication, confidence and acceptance of food derived using biotechnology. To this end:

Meet consumer needs with enhanced quality and safety of products.

Consult the public early in the research and production process to determine consumer acceptance of products.

Use biotechnology to assure food safety and quality.

Enhance the credibility of the regulatory system.

Expand dialog with the public and in K–12 and college classrooms.

Discuss the results of safety and socioeconomic research publicly.

ANIMAL GROWTH PROMOTANTS

While workshop participants identified several important issues and goals, there was a divergence of opinion on how they might be addressed. The following recommendations were presented:

Assess the broader social impact of animal growth promotants (and all new technologies). There was disagreement as to how, what, by whom and when. This was the most controversial issue discussed.

Assess the process by which AGPs are regulated for human, animal and environmental safety for possible improvement. The methodology proposed to accomplish this goal includes: 1– identifying the concerns of the different constituencies, possibly in a meeting or via a survey; 2– briefing these groups on the regulatory process as they each may not have a complete overview of the process; 3– developing a white paper that reflects the concerns of the groups with subsequent feedback in order to reach a consensus; and 4– developing solutions for the concerns identified.

Make information available on the risks and benefits of AGPs. The methodology proposed included: 1– research to determine potential risks and benefits; 2– public education to disseminate this information; and 3– public policy set to insure that relevant, credible, objective, non-proprietary information would be available to all.

Eliminate international regulatory disparities with a goal of (free/fair) trade. To accomplish this: 1– establish an international dispute settlement mechanism such as GATT; 2– conduct research in a neutral, independent setting such as an international research center; establish objective international standards such as CODEX; and 3– educate consumers possibly using UN/FAO funds.

TYING IT ALL TOGETHER

As I reflect aloud for the next few minutes on my impressions of the meeting that we have just been through together, I ask that you reflect on your impressions as well. Obviously, our reactions to the meeting come out of the experience and perspective that we bring to the meeting and your reactions will be somewhat different from mine.

PATRICIA B. SWAN

Interim Provost
211 Beardshear Hall
Iowa State University
Ames, IA 50011

Swan serves on the advisory boards of several national institutes and governmental entities. With graduate degrees in biochemistry and nutrition, she has researched a wide variety of topics related to metabolism and cellular function.

I want to do two things. First, to review the theme of the meeting, *agricultural biotechnology at the crossroads* and use that to talk about what we have heard and said together; and second, to spend a few minutes reflecting on the actual way in which we met together and the quality of our communication.

The call to the meeting elaborated the theme of the meeting. To remind you, "In the late 1970s, several published reports emerged extolling the benefits of biotechnology, predicting the outcome of exciting new research agenda, anticipating the potentially vast benefits...Biotechnology became the byword of the 1980s, with new techniques for mediating the determinants of heredity, almost anything seemed possible. These new techniques were expected to raise agricultural productivity while reducing chemical use in agriculture, thus leading to a cleaner, healthier environment. These predictions reflected the tenor of the times, in which biotechnology was widely held to be a technology that would revolutionize the coming decades." Not surprisingly, this point of view was expressed optimistically. It was a critical juncture, I would assert, a time when we were trying to interest investors of all types in the promise of these new technologies. We (scientists) were going to Congress and to the public, and quite unusually, to venture capitalists. The science of molecular biology was giving rise to these new technologies but the question remained "Could the resources be found to develop these technologies?" Promises were made at this juncture and like many promises, they sometimes come back to haunt us.

I am reminded of a story Garrison Keillor told the other day. He was talking about a particularly life threatening critical juncture and he found himself beginning to pray. "God, if you will just help me out of this problem, *alive*, then remember those promises I made to you back in Chicago at O'Hare when the landing gear didn't go down on the airplane...?" Thus the old promises came back to haunt him.

One speaker, Roger Salquist, disputed the premise that the promises have failed to develop as predicted. Instead, he asserted that biotechnology to date has met the knowledgeable predictions that were made 10 years ago and that we are now on the verge of having an impressive portfolio of products.

The call to the meeting went on to say, "Clearly, complex, unforeseeable biological, institutional and socioeconomic changes have emerged that place biotechnology in a new context. Today, constraints and incentives create an environment for research and application that is very different from the rules and regulations perceived to constrain both scientific and public policy development ten years ago. This Third Annual Meeting will assess the reasons why many biotechnology innovations have failed to develop as predicted, identify and evaluate the constraints and incentives that currently drive research, commercialization and acceptance of biotechnology products and seek ways to assess their appropriateness or inappropriateness."

Appropriate to the theme of the meeting, several crossroads have been set before us. As we meet these crossroads, the question is "Will the direction we choose give fulfillment to the promise that biotechnology will indeed revolutionize the era ahead?" or, "As we meet these crossroads will we find that we were wrong about the promises of biotechnology?" During the last two days, several types of crossroads were discussed, not just the major one suggested by the call to the meeting, but several smaller crossroads within that. Some of the larger intersections are behind us. It is always good to reflect on the roads crossed in the past and the decision made to take a certain direction. I would assert that we have already crossed at least three crossroads.

CROSSROADS PASSED

Ten years ago, there were many different definitions of biotechnology. The word was used to signify different kinds of activities and enterprises and was used in many different ways. During the last two days, several people offered definitions as if they still wanted to clarify, but their definitions agreed. There was a consensus that biotechnology is a set of technologies, a set of tools, a set of methods and not a monolithic entity.

Another crossroad we have passed is the question of whether or not these technologies will be used in agricultural science. We have lots of evidence that we are beyond that question—they are being used now and they will be used in the future.

A third crossroad that we have passed is the question of whether or not to regulate the products of biotechnology. Ten years ago, some people said "They are like natural products, don't bother to regulate." and there were others who said, "They are VERY unnatural products, regulate everything about biotechnology...the process, all the products, etc.". We have come to that crossroad and for a variety of reasons are agreeing that we will regulate only the products of biotechnology. No longer is there any sizeable voice suggesting no regulation at all.

FUTURE CROSSROADS

There are several crossroads we still face. We have come upon them, but we have yet to make decisions on which way to go. Over the last two days, a major crossroad permeating much of the discussion at all levels from the general to the highly specific was the question of whether biotechnology will drive our future. Are we going to be led by the technology, caught up in it, or carried along by it? Or, as some of our speakers have asserted, will other forces set the direction for us and determine the way in which we use these technologies that we call "biotech"? Which will be the driving force?

Forces proposed were economic forces of varying types, including international trade; the risk-benefit issues for farmers, consumers, etc.; and the developing and overriding environmental concerns in our society today. The question is "What other factors will be allowed to drive biotechnology if we do not allow the technology to determine our future?" There is no agreement yet. Some would say the technology should have free reign, while others would say that economics should be the overriding concern. Still others would say there are social concerns which must then modify our economic system. It seems there is considerable disagreement arising from the differing values we bring to the table, as to what the role should be of these other forces which may determine the future use of biotechnology.

There are other crossroads which received attention. I had not predicted that regulation would be as pervasive an issue as it was in the last two days. Throughout the workshops, as well as in the informal discussions here together, regulation seemed a compelling topic. I heard agreement on the point that regulations should be fair, timely and there should be a great deal of public trust in it. I heard some agreement that regulation is not perfect yet. I heard much disagreement as to its points of imperfection. Clearly we have had too little dialog on how to regulate, who regulates whom and what. I found a lot of consensus that we should have some regulatory activity to assure safety. We do not seem to agree whether efficacy is an important point in regulation. Currently some products in our market are judged by whether or not they are efficacious, while others are not. Where biotechnology products should lie within the whole regulatory scheme is a point over which we have yet to reach a consensus.

We also disagreed on how well we are doing—I heard one speaker say that he thought there was considerable room for improvement and he focused on the complexity and uncertainty on the multitude of organizations involved in regulation. Another speaker said he thought it was going very well – that his own experience with the regulatory process had been extremely smooth, processing had gone in a timely fashion and that he was anticipating nothing but the best. In the unknown time ahead we will look at the regulation of actual products coming to market, commercialization as opposed to field trials and testing stages of development of products. The commercialization of products will precipitate many difficult regulatory decisions.

In sum, I heard more concerns about points of imperfection in the regulatory system than I heard agreement that it was doing well. Although a lot of regulatory sub–issues were explored in the workshop reports, it would be redundant to summarize them again this morning.

Let me comment on a couple of other areas where I think I heard some agreement. They are old questions. Should we be regulating the science that is behind these new technologies? There was not quite 100 percent agreement, but a fair consensus that we should not regulate the science *per se.* We should regulate the products of the technology arising from the science, and while we agreed to regulate the products, we disagreed as to the criteria we should use. The unmapped territory ahead has given rise to a lot of uncertainty and feelings of insecurity, heightening the need for more exploration of what it is that we are going to be doing in the process of regulating products as they are commercialized.

Funding was another crossroad that came up throughout the meeting. It was amazing to me that it came up as frequently as it did. Although funding is part of my job, a part of my business, I had not imagined that funding could be so broad a concern as it appears to be in this meeting. Dollars. Whose dollars? There was much discussion about the role of public funding and private funding. The kinds of activities which face us in developing products from biotechnology give rise to the questions, "What should be the role of public funding? What should be the role of private funding? Do they have different roles?" This is not a discussion limited to biotechnology, but the kind of questions that arose suggested that we need much more discussion of funding, especially since there is not an understanding among us about what our position is on the roles of public and private funding. There are many questions about the appropriateness of different kinds of work using public or private funds, as well as questions about how these funds are currently being used. Again there is room for a lot more discussion and a chance to get much more highly specific about what it is that concerns us in these arenas.

Funding for commercialization also came up. I was surprised that it did not come up more often. I had predicted that it would come up more because the press is writing a lot about where we are with regard to funding for biotechnology companies, both biomedical and agricultural biotechnology.

There was a lot of discussion about the myriad lists of critical technologies that are emerging from Washington and related groups; discussion about the current state of competitiveness in biotechnology and what we can expect in the future. Those emerging lists are saying that we are currently very competitive in biotechnology in the United States as compared to the rest of the world, but that funding concerns, especially for commercialization of products, as well as regulatory and other concerns, are giving rise to questions about how long we will remain competitive. These are all popular issues in the press today; however, there was not as much about these issues at our meeting as I had anticipated. "How critical are these issues?" presented another crossroad on which we expect to differ.

Other crossroads...what type of research should we be doing? Who decides? "*Who* decides" kept coming up and not getting answered, except "ME, I'll decide." Whoever is talking is the person who wants to decide. It is an important issue, one needing some public understanding and consensus. What decisions are appropriate in the public arena? What are appropriate in the scientific arena? How do those two inform each other as they each make their appropriate decision? Again, we have not had enough dialog on these issues because I heard more questions than I heard any attempts to provide answers or even points of view.

The technologies we have been discussing for the last two days are developing extremely rapidly. We are making remarkable progress in biotechnology, but it may be that the technology has outstripped our basic knowledge with regard to biological systems. What limits us in our ability not only to use these technologies, but to use them wisely, is the fact that we do not have enough knowledge of the biological systems in which we are using them. This limitation came up in many different ways during the meeting. There was a strong call for getting on with the business of understanding these biological systems more clearly so that we can be more thoughtful about what we are doing with technology. There seemed to be a pervasive concern that we have outstripped our ability to assess impacts of the technology. We are not skilled at assessing those impacts. We all say we worry about them, and I think we do. We agree that we worry about the social impacts, economic impacts and other kinds of unforeseen consequences of these new technologies. We are uneasy because we do not know if we can successfully assess the future. We are uneasy because we do not know if we will take the time to do it, and we do not know who will do it. These are points about which we have a great deal of unease. From discussions, it seemed there was more concern and unease than disagreement or conflict. We do not yet know where we each have positioned ourselves within this whole arena. Thus there also looms a crossroads with regard to our research agenda, not only funding or the type of research we are doing, but also what we need to do in order to make ourselves better able to assess the impact of these technologies.

Communication was another pervasive concern, but I do not know if "communication" could ever be considered a crossroads. We almost always refer to it as a two-way street or many two-way streets. Communication is certainly something that everyone was worrying about. We agreed that we wanted to communicate. We agreed that we wanted to communicate *better.* We seem to agree that our communication is not as good as it should be. It is typical of society today that we have this concern. Unfortunately, we did not have the time to explore how we could improve communication. What could we do, specifically, to improve our communications, improve it in the regulatory arena, improve it in the research–decision–making arena, improve it in the funding arena? There were many places where we asked for improvement, but did not really explore what we were doing wrong or not enough of, or should be doing more of, to improve communication. The issue of education came up frequently... education that would provide a more common understanding on which to base our communication and education to improve decision making. There was a strong call for more exchange of information, an open exchange of information and lots more of it.

HOW DID WE DO?

Those are some of the crossroads that I heard being discussed in the last two and half days. How did we discuss them? NABC has the goal of bringing together diverse views, openly discussing and examining those diverse views and in Ralph Hardy's words, "We're coming together to speak, to listen, to learn." How well did we do that? I offer special thanks to the NABC/Joyce Fellows, because I asked them at some length last night how well they felt we had achieved this goal. They had a fairly optimistic answer for me. They seemed to feel that we had a good diversity of views. Here they mentioned two perspectives which they were not sure were represented as well as they should be. Regulators were one group. Another group whose view they thought was missing was the farmers who will actually employ these technologies. But they felt that the discussion had been open and that there had been an exchange of the many views present here. On those grounds, they were rather optimistic about the goal of the meeting being achieved.

I always ask, "How vigorous and how fearless was our examination of issues?" Those of you associated with the University of Wisconsin know about one of the formative statements behind the University which has touched all of us at UW. It is a statement on a plaque in the main administration hall there. It comes from a statement made by the Wisconsin Board of Regents before the beginning of this century: "In all lines of academic investigation, it is of the utmost importance that the investigators should be absolutely free to follow the indication of truth, wherever they may lead. Whatever may be the limitations which trammel inquiry elsewhere, we believe that the great state University of Wisconsin should ever encourage that continual and fearless sifting and winnowing by which alone the truth can be found." I heard some fairly courageous sifting and winnowing in the last two days, although I am not sure it measured up to the standards set by Wisconsin's Regents 100 years ago.

Perhaps we were not completely fearless. We may not have been willing to voice our opinions as strongly and with the conviction that is required to really examine perspectives and to push to a true consensus. We can excuse ourselves, for in two days what can you do? We can excuse ourselves because we are all nice people and we do not want to hurt other people's feelings. We can excuse ourselves for a variety of reasons. However, from a professorial tradition, I would have to say in that regard that I think the best grade we could give ourselves would be a B.

POST MEETING SUMMARY

JUNE FESSENDEN MACDONALD
Deputy Director
NABC
159 Biotechnology Bldg.
Cornell University
Ithaca, NY 14853

In a post-meeting session, members of the NABC Council and Operating Committee, NABC/Joyce Fellows and Graduate Fellows, speakers, workshop chairs and session moderators came together to discuss the meeting and its recurrent themes and to develop a follow-up strategy for NABC and its member institutions.

Following lively debate, three priority action areas were identified:

1 – the need for adequate funding to further develop our national research capacity with strong support for stepwise increases over 5 years to full funding of $500 million for the National Initiative for Research on Agriculture, Food and Environment.

2–the need to include a more substantive social and economic impact assessment in all technology impact assessments of emerging agricultural biotechnology applications.

3 – concern about the need for additional regulatory definitions as well as what the appropriate level of regulation by the federal government should be and the role of individual states.

As an initial step, letters prepared with input from designated attendees were sent from NABC to appropriate Congressional members and agency officials. The strength of these recommendations lies in the backing they have received from participants with different interests and agendas who came together at NABC3 to promote current, specific needs in agricultural biotechnology.

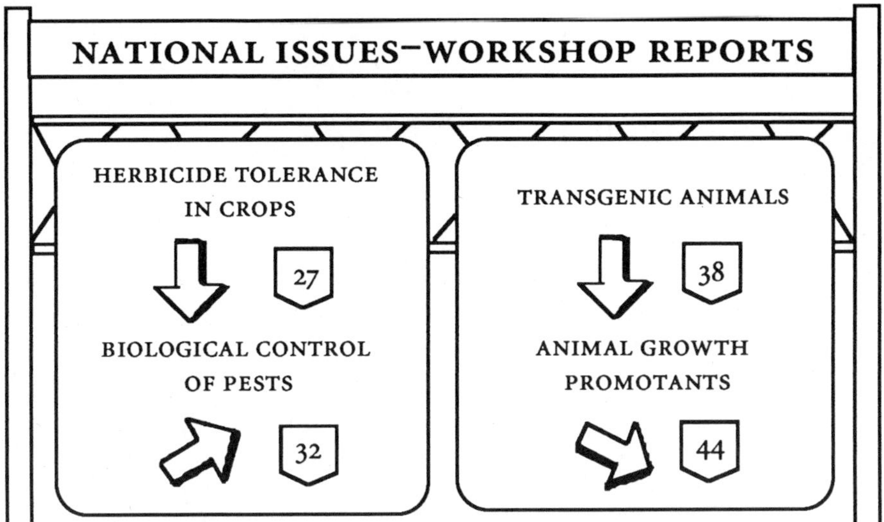
NATIONAL ISSUES—WORKSHOP REPORTS
HERBICIDE TOLERANCE IN CROPS
27
BIOLOGICAL CONTROL OF PESTS
32
TRANSGENIC ANIMALS
38
ANIMAL GROWTH PROMOTANTS
44

c. 1954

WORKSHOP REPORT
HERBICIDE TOLERANCE IN CROPS

Preparation of NABC for the workshop on herbicide tolerant crops (HTCs) began with a meeting at Iowa State University (ISU) during October 1990. Participants were invited from academia, federal and state government and other selected organizations to conduct a benefit/risk assessment for the introduction of herbicide tolerant crops in Iowa. The report of that meeting served as a background paper for this workshop. See page 179.

CO-CHAIRS

WALTER R. FEHR

Director of Biotechnology
Iowa State University
Ames, IA 50011

REBECCA GOLDBERG

Biologist, Environmental Defense Fund
New York, NY 10010

NABC/JOYCE FELLOW

JEFFREY C. FEARN

Plant Production
Boyce Thompson Inst.
Ithaca, NY 14853

This workshop differed from the one at ISU in three important aspects: 1 – Only a few of the participants in Sacramento had attended the meeting at ISU. Consequently, the workshop at NABC3 was essentially an independent evaluation of HTCs. 2 – There was considerably less time available at NABC than at ISU, which restricted the topics that could be discussed. 3 – The participants at NABC chose to devote a large fraction of their time to a discussion of appropriate government oversight of HTCs.

The first step in the workshop process was to list comments that one or more attendees considered to be important with regard to HTCs. The wide array of comments reflected the broad range of interests and opinions among the attendees.

The second step was to select from the list a limited number of the issues. Eight major issues were identified and the attendees individually ranked their importance. The eight issues are listed below in their order of interest to the group as determined by vote of the attendees.

1–Will HTCs modify herbicide use in agriculture and forestry?

2–Risk assessment relative to human and environmental concerns.

3–Socioeconomic impacts.
4–How will HTCs be regulated?
5–Public perception of HTCs and responsibility for HTC education and appropriate use.
6–Use of public research funds to develop HTCs.
7–HTCs affect on food safety.
8–Affect of HTCs on crop management by farmers.

The first four issues were clearly the most popular; therefore, with the limited time available, the attendees decided to consider only those four issues. The attendees were divided by individual preference into four groups and each group discussed one of the topics. Each group was asked to a) clearly describe the issue, b) define the goal that the group wanted to achieve relative to the issue, c) provide recommendations on how to achieve the goal, and d) describe the criteria by which progress toward the goal would be evaluated. Summaries of the reports from the four groups are provided.

Issue 1: There is controversy about how HTCs will modify herbicide use. Despite the controversy, there was general agreement that the primary goal of research should be to ensure that HTCs result in the safer use of safer herbicides. It was also considered important to ensure that the use of HTCs should be compatible with integrated pest management systems of weed control. It was also considered important to ensure that the use of HTCs does not divert efforts away from integrated pest management systems of weed control. With new technologies becoming available to farmers, only one of which is HTCs, farmers must learn how to integrate the various options into the best management plan for their farm. Optimum weed management strategies should rely on integrated approach, which may include crop rotation, cultivation and the minimum use of chemicals.

To attain the goals defined for HTC development, it was recommended that *field research studies should be conducted that compare alternative weed control strategies.* The studies should consider how HTCs will affect the overall production system, including soil conservation practices and minimum tillage systems.

It is recommended that *HTC research should be focused on herbicides with the following attributes: low toxicity to non–target species, including humans and wildlife; low residues in the environment, including ground wa-*

ter, surface water and air; nontoxic residues in the crop and food; low use rates; appropriate degradation (breakdown) to benign breakdown products; cost effectiveness; compatibility with alternative weed management strategies; compatible with technology improvements in the way the herbicide is applied; and increased reliability of weed control accompanied by improved crop yields. It will be important to investigate whether the use of certain HTCs would increase worker or consumer exposure to a herbicide. The group did not discuss whether research on HTCs should occur in the public as well as the private sector.

Success in achieving the goals of the safer use of safer herbicides and ensuring that HTCs are compatible with integrated weed control systems would be measured by the identification and development of economical weed control systems for farmers that also are beneficial for society.

Issue 2: There is a need to identify any health and environmental risks specifically associated with HTCs.

Issue 3: The socioeconomic impact of HTCs should not be a regulatory criterion for product approval by government agencies. Information concerning the socioeconomic impact of HTCs should be publicized to permit consumer choice and any requisite governmental mitigating effort.

Issue 4: There is a lack of regulatory policy to regulate some aspects of HTCs.

The reports of the groups that considered issues 2, 3 and 4 are presented collectively, since all three issues concerned the regulatory process. Participants first agreed that any health and environmental risks associated with HTCs need to be identified and that criteria to assess these risks should be developed. Examples of possible risks include the transfer via pollination of a gene conferring herbicide tolerance from a genetically engineered crop to a related weed and the safety as food of crops genetically modified to tolerate herbicides.

The group made three recommendations concerning possible health and environmental risks. First, *environmental data for HTCs should be evaluated on a case-by-case basis.* The biology of the crop, the nature of the genetic modification and the characteristics of the herbicide the crop has been genetically modified to tolerate should form the basis for evaluations. Such evaluations should be performed in a reasonable time frame. Second, *the federal government should, on an ongoing basis, use public funds to construct a database from information obtained from small-scale field tri-*

als of HTCs. The database would be used to codify data relevant to questions concerning environmental risks and to identify knowledge gaps. Third, *guidance should be provided to ensure that seed companies and other institutions developing HTCs ensure the food safety and quality of the crops.* Some participants cited guidelines developed by the International Food Biotechnology Council[1] as appropriate.

For both the first and third recommendations, participants felt that the impact on the cost of products and the development of minor versus major crops should be considered.

Participants agreed that regulatory policy covering some aspects of HTCs is lacking. Although the group did not discuss precisely which aspects of regulatory policy needs to be clarified or established, the group agreed that a fair, timely process clarifying or establishing regulatory policy is needed. Participants recommended that *a policy stating what is regulated and who regulates it should be articulated by January 1, 1992.* (This date may seem ambitious, but many participants felt a pressing need for regulatory policy.) Within a reasonable time after a regulatory policy has been articulated, the regulatory process should be "tested" with a specific HTC which is ready to go through such a process.

The group set three general criteria for such a regulatory process. First, the regulatory process should provide meaningful opportunities for public input. Second, the regulatory process should allow industry to proceed in a fair and timely manner. Third, as appropriate, responsible health and environmental safety reviews should be conducted under the regulatory process.

Participants also agreed that the *socioeconomic impacts of herbicide tolerant crops should not be a "fourth criterion" for government regulatory approval for product commercialization,* as has been proposed in the European Community. Products should be judged on the first three criteria: human and environmental safety, quality and efficacy.

The group made two recommendations concerning socioeconomic impacts of HTCs. First, *government institutions that already deal with socio-*

---

[1] International Food Biotechnology Council. *Biotechnologies and Food: Assuring the Safety of Foods Produced by Genetic Modification.* Printed as a supplement to *Regulatory Toxicology and Pharmacology,* Volume 12, December 1990.

economic concerns may wish to consider the socioeconomic impacts of herbicide tolerant crops and if appropriate, mitigate any impacts. (Examples of mitigating actions currently employed by the government range from providing unemployment compensation to subsidies for new enterprises in adversely affected communities to antitrust actions that reduce or redistribute market shares.) Second, *public discussion of the impacts of HTCs, including their socioeconomic impacts, should be fostered.*

WORKSHOP REPORT
BIOLOGICAL CONTROL OF PESTS

Biological control methods have significant potential to become important tools for managing populations of agricultural pests. However, in spite of over 100 years of use in the United States, biological control has made little impact as a viable alternative strategy for pest control. It is clear that biological control has not been a significant national priority and there is no coordinated national program in place to enhance its development and implementation. The main objectives of the workshop were to develop strategies for a national effort on *how to make biocontrol work* and to propose recommendations for implementation of these strategies.

CO-CHAIRS

ROBERT R. GRANADOS

Research Director
Plant Protection
Boyce Thompson Institute
Ithaca, NY

MAUREEN K. HINKLE

Director of Biotechnology Policy
National Audubon Society
Washington, DC

NABC/JOYCE GRADUATE FELLOW

JOHAN F. SWINNEN

Agricultural Economics
Cornell University
Ithaca, NY 14853

The workshop brought together approximately 45 individuals with different backgrounds: industrial, environmental, and consumer groups, universities and regulatory agencies. The participants were asked to project their vision of how they thought biological control should be part of our agricultural future. From these discussions, some broad perspectives emerged. The most important broadly-shared vision was the hope that the use of biological control in pest management would increase significantly in the coming years and reduce our dependence on chemical pesticides. Furthermore, several individuals mentioned the importance of the use of biological control as part of an integrated pest management system. Most of the individual visions that did not explicitly mention this broad perspective focused on issues and problems that hitherto have prevented biological control from becoming an important part of pest control.

While biological control practices offer an environmentally-friendly alternative to chemical pesticide use, numerous constraints have prevented bio-based technologies from becoming an important means of pest management. The participants identified four major areas of concern within biological control that deserved separate attention: 1) technology, 2) regulation, 3) commercialization and 4) adoption. In addition, it was agreed that there is an overall lack of advocacy to get biological control on the national agenda.

The group was divided into four subgroups, each focusing on one of these areas of concern. Each subgroup was charged with developing goals in the different areas and recommendations to achieve those goals.

TECHNOLOGY

Important technological issues are:

— the lack of sufficient knowledge of the potential for biological control to be commercially viable (i.e. efficacy, market potential, production technology, etc.);
— the lack of strategies and fundamental knowledge underlying the productive use of biocontrol mechanisms; and
— insufficient funding to overcome these hurdles.

There was general agreement that there was insufficient knowledge on mechanisms of biological control at the genetic, molecular, population and ecosystem levels. The goals set on the technological level were to provide sufficient funding and/or develop incentive schemes that would encourage research on the viability of biological control for the grower and commercial producer and on existing defense mechanisms in nature.

Recommendation T-1: Establish and maintain basic and applied research programs to address scientific issues.

These research programs should focus on the following areas (in order of priority): 1) host-pest-biocontrol agent interactions, 2) ecological relationships between the target pest, its environment, and biological control agent, 3) host resistance mechanisms and 4) compatibility of biological control agents with chemical agents in integrated pest management (IPM) programs.

Recommendation T-2: Provide public funding and incentives for 'public good' (commercial products) types of biological control.

Recommendation T-3: Provide incentives for 'private good' types of biological control.

Various options that were mentioned in the workshop discussions, without necessarily being agreed upon by the group as a whole, included diversion of money from the Clean Water Act and establishment of a pesticide tax. These revenues would be used for field demonstrations of biological control. For 'private good' incentives the following ideas were proposed: an 'Orphan Drug Act' for small market biopesticides, research and development tax credits, clear and concise regulations for field testing and registration of biocontrol agents, and lowering capital gains taxes to help research and development investments.

REGULATION

The following goals should be set in developing the regulations for biological control agents:

— establish clear-cut regulatory procedures for all biological control agents;

— establish risk-benefit based regulation; and

— establish guidelines permitting exemptions for interstate movement of cultures of biological control agents for research purposes.

Therefore, the following regulatory recommendations were forwarded by the group and listed in order of priority:

Recommendation R-1: Redefine regulatory procedures by: 1) defining agency responsibility for organism groups; 2) defining criteria/characteristics representing risks and benefits; and 3) establishing a fixed time for regulatory decisions on field test applications.

Recommendation R-2: Improve communication by: 1) facilitating access to federal/state regulatory procedures; and 2) establishing voluntary mechanisms to share results of safety tests between investigators and between federal and state authorities (such as the National Biological Impact Assessment Program data base in the USDA).

COMMERCIALIZATION

There are three major impediments to the commercialization of biological control technologies:

— biological control agents are not sufficiently reliable and available for commercialization;
— incentives to commercialize the biological control products are not sufficient; and
— product cosmetic quality standards of the consumer are too high.

Therefore, incentives to increase the use of biologicals and minimize these barriers are necessary. A number of policy options were discussed in the group and it was agreed that they should be evaluated on criteria such as their effectiveness in achieving the goals, their ease of adoption, their political feasibility, their cost and their impact on the international competitive position of the U.S. farmers. From this discussion, the following recommendations were adopted:

Recommendation C-1: *Set up national research centers to develop biological control methods with local/cooperative, clearing houses for basic and applied information on and the delivery of biological control agents.*

Recommendation C-2: *Modify crop support programs to encourage diversification and provide insurance premiums against crop loss to farmers who use biological control agents.*

Recommendation C-3: *Develop programs to change the lack of tolerance for imperfections at all levels of the marketing systems.*

ADOPTION AND IMPLEMENTATION

The most important issues needed to be addressed in this area were:
— The lack of information on biological control strategies. Potential users do not understand which alternative options exist — or could become available — and how they should use them.
— Economic constraints. Existing biological control practices often have a cost disadvantage.
— Distributional problems. The overall access to and availability of the agents is limited.

Several initiatives were recommended that could overcome these problems. The first initiatives focused on transmitting information on alternative biocontrol technologies. An important criterion in setting up these programs should be the ability to demonstrate the efficacy of the biocontrol technologies.

A second group of initiatives would enhance the access to biological technologies and their availability. For this purpose, marketing and distribution systems should be developed. It was suggested tying this into established IPM programs.

Finally, it was argued that incentive mechanisms should be developed to make biological control technologies more cost effective. Suggestions were: 1) to set up 'shared risk programs', which would reduce the increase in production risk that the farmer incurs by using biocontrol agents 2) to provide inexpensive insurance for programs under testing and evaluation and 3) to tax chemical usage to reflect its total social cost and use these revenues to support biocontrol adoption programs.

The selection of these programs should be based on a thorough evaluation of the environmental costs of the technologies they promote and on their financial sustainability, both from a government (taxpayer's) perspective and from the producer and farmer's point of view.

The group agreed on the following recommendations:

Recommendation A-1: *Establish a national program for the promotion of biological control. This included: 1) developing educational and informational materials and their distribution and 2) establishing demonstration projects on farms.*

Recommendation A-2: *Establish programs to reduce the initial cost disadvantage of biological control agents over traditional pest management techniques. This could include taxing pesticides/users and use of the revenues for these programs.*

CONCLUDING REMARKS

Four issues emerged throughout the subgroups as important for the development and increase in the application of biological control in the coming years. The first was the *lack of leadership* in the area of biological control as a whole. It was the feeling of the workshop attendees that biological control has to get to the national agenda and, for this to happen, advocacy is necessary from academicians, legislators, and other public policy makers. Secondly, everyone agreed on the need for *effective and reasonable regulatory procedures.* Third, the establishment of a *tax on pesticides* to provide government revenues for the promotion of biological control applications was raised in all subgroups. Finally, there was a strong need identified for an increase in *government funding* in the public arena (research and extension) and the establishment of *(financial) incentives* for an increase in supply (industry) and demand (farmers) of biological control agents.

A number of problems were not addressed in the short time the workshop group was together; and several issues were deferred to a "debate continues" category. Among these were the questions on how to manage resistance to biological control agents, approaches and problems associated with narrow genetic resources, the definition of biological control, and patentability of biological control agents.

WORKSHOP REPORT
TRANSGENIC ANIMALS

The possibility of transferring sections of genetic code into the genome of an animal—creating thereby a new genetic resource for a species—has been among the most exciting and dramatic applications of biotechnology. To date, there have only been a few laboratories attempting to apply transgenic technology to farm animals. Although commercial application of transgenic animal technology for enhancing the quality or quantity of the food supply is not imminent, it is not too early to begin discussing the social and ethical issues associated with transgenic animals.

CO-CHAIRS

JAMES MURRAY

Animal Science
UC Davis, Davis, CA 95616

PAUL B. THOMPSON

Director, Center for Biotechnology Policy & Ethics
Texas A&M University
College Sta., TX 77843-4355

NABC/JOYCE FELLOW

ROLEY PIGGOTT

Agricultural Economics
UC Davis, Davis, CA 95616

The workshop on transgenic animals included representatives from public and private research organizations that are either conducting or contemplating research on transgenic animals, representatives from several public interest groups, and several individuals interested in the politics and policy issues associated with biotechnology. There were not participants whose interest was defined in terms of opposition to transgenic animals, nor were there participants who expressed concerns that would rule out research and development of transgenic animals. This fact is important in evaluating the consensus statements that are summarized below because they do not reflect the view of the groups that would be most likely to oppose the development of transgenic animals.

Participants in the workshop on transgenic animals listed several dozen topics for discussion. These topics were summarized into the following four headings:

1) Anticipation and management of unintended consequences

2) Socioeconomic impacts

3) Funding–priorities and participation
4) Food safety and consumer acceptance

Participants divided into four subgroups to work on these areas. Each subgroup was charged with constructing a statement of the issue, defining the goal sought, and identifying strategies for accomplishing the goal. Subgroup reports were reviewed by the entire workshop.

In general, these discussions did not generate a high level of controversy or emotional intensity with respect to transgenic animal research. The workshop group felt that, since transgenic research is in the early stages, it would be some time before issues associated with the development and application of transgenic animals would become public. The group agreed broadly, at this juncture, that more basic research on the techniques and potential of transgenic animals in agriculture would be required before it would be possible to conduct well informed discussions of the social, ethical and policy issues that might one day be associated with transgenic animals.

In the following workshop summary, each of the four areas is discussed, with general conclusions drawn in the final section.

UNINTENDED CONSEQUENCES

Issue—Unintended and unwanted consequences of technology are most often the source of social and ethical problems. Aside from food safety and socioeconomic dislocation, both discussed below, transgenic animals may be associated with several broad categories of unintended consequences. First, there is the possibility of environmental risk, including impact upon genetic diversity, animals as disease vectors and effects upon wild populations, as well as pollution that may be associated with animal production. Second, there are issues associated with the well–being of the animals themselves, both in terms of their health and in terms of their ability to lead relatively tranquil lives. Third, there is the possibility of consequences not readily anticipated.

Goal—Animal genetic research should be conducted to enhance human and animal well–being at an acceptable risk to animals, humans and the environment.

Strategies—The workshop discussed three strategies expected to help achieve this goal.

1) *Adequate funding.* If there are sufficient public funds to conduct research in the area of transgenics, the group felt that many of the worst problems of unintended consequences could be avoided by reducing the pressure to rush to market and by supporting research on risk, including frames of reference.
2) *Responsible science.* The second strategy stresses use of peer review panels, such as current animal–care committees, to assure that research on transgenics follows existing rules both for containment of recombinant DNA experiments and for animal welfare. This strategy requires open disclosure of the peer review and enforcement process. It will succeed only to the extent that the public is aware that these procedures are in place, and has the opportunity to express concerns to the appropriate bodies.
3) *Quality science.* The third strategy is to bring scientific resources to bear upon anticipating and assessing unwanted consequences. Both unwanted and beneficial outcomes can be anticipated through the development of model systems. For this strategy to succeed, there is a current need to place transgenic research within the framework of basic, rather than applied, research. Once models have been developed, it will be possible to conduct cost/benefit assessment of future transgenic research.

SOCIOECONOMIC CONCERNS

Issue—Can transgenic animals be used to enhance the quantity and quality of animal products in an efficient, sustianable manner (i.e., economically, environmentally and socially sound)?

Goal—To minimize the negative and maximize the positive socioeconomic impacts of transgenic animals and their products at the local, national and international levels.

Strategies—There are four key steps that must be completed to achieve the goal.

1) Determine the type of transgenic animal/product, the time frame in which it will be developed and adopted, and where and how it will be used.
2) Identify direct economic consequences (good and bad) of adopting transgenic animals/products. It will be important to include second

and third order consequences, i.e., downstream effects of transgenic animals.

3) Identify possible environmental benefits/costs associated with the use of transgenic animals/products.
4) Identify segments of world society that are affected by transgenic animals technology. This step includes involving society, developing public policies and conducting research on the socioeconomic impacts of transgenic animal technology.

FUNDING

Issue—Transgenic technology is merely a research and technological tool. As such, it should not be singled out as a special priority for funding. Rather, basic animal genetic research funding should be increased, because an increased knowledge of animal genetics will help increase the welfare of humans and animals.

Goal—Public funds should be used to increase the genetic knowledge base.

Strategies—There are three separate, but mutually inclusive, ways to go about securing this goal.

1) Preferably, funds will be distributed by peer review through competitive grants with public participation during the decision–making process.
2) Fund secondary and undergraduate education in animal genetics.
3) Public policy should encourage industrial funding for animal genetic research, with public participation and oversight.

FOOD SAFETY AND CONSUMER ACCEPTANCE

Issue—How can modern animal genetics be used to enhance the quality and safety of food, while minimizing the possibility of unanticipated deleterious effects? The existing research and regulatory systems should be more effective in building public communication, confidence and acceptance of food derived using biotechnology.

Goal—Ensure the safety and quality of food through modern genetics and develop (public) consumer communication that assures decisions are made on an informed basis.

Strategies for assuring quality and safety—

1) Meet consumer needs with enhanced quality and safety of the product. It will be necessary to produce products through modern animal genetics that are attractive and desirable to consumers, and to create a "market pull" for new products.
2) Conduct research to avoid unexpected consequences affecting food safety. It is important to clearly and demonstrably conduct research in parallel that looks for and avoids unexpected safety or quality problems with animal products derived through modern animal genetics.
3) Use biotechnology to assure food safety and quality. It will soon be possible to use the modern tools of biotechnology, for example for microbial probes contamination, to enhance the safety and quality assurance programs in food inspection.
4) Include transgenic animals in the broader context of animal molecular genetics. Since the application of transgenic animals may raise concerns, the subject should be discussed as part of the overall program of research and the products resulting from the use of modern animal genetics; transgenic animals are only one of the tools used to achieve the desired new products.

Strategies for improving consumer information—

1) Enhance the credibility of the regulatory system. Trust in the regulatory system to assure a safe and wholesome food supply has eroded. Research and changes in policy should be explored to improve the performance and image of the regulatory system.
2) Expand dialog with the public, including consumers, K–12 and college. The public image of the food supply is greatly influenced by what children learn in the classroom and what the consuming public is exposed to in the media. There should be a balanced effort to expand two–way communication with the public regarding concerns about the safety and wholesomeness of new products and processes from transgenic animals.
3) Early in the research and technology to produce products from transgenic animals, the public should be consulted to determine consumer acceptance of the products.
4) The results of safety and socioeconomic research should be discussed publicly. Elsewhere in this workshop, participants emphasized the

need to conduct parallel research to minimize socioeconomic impacts of the results of research on animal molecular genetics. A corollary to this issue is to conduct parallel research to assure that unanticipated results are minimized. The public should be made aware of these efforts as a part of the overall communication process.

CONCLUSIONS

In general, workshop participants concluded that public issues associated with transgenics were not urgent, primarily because applications of transgenic technologies as they affect agriculture and the food supply seem remote at the present time. It was expected to become more prominent. In this connection, it will be important to determine whether existing guidelines for treatment of research animals are adequate to assure the well–being of transgenic animals.

WORKSHOP REPORT
ANIMAL GROWTH PROMOTANTS

The following four issues evolved from the workshop on animal growth promotants (AGPs). There was consensus in identifying the important issues and goals with respect to AGPs. Divergence of opinion occurred in how they might be addressed. The order of the goals does not signify their importance. No discussion as to the relative rank or importance of each occurred.

CO-CHAIRS:

FLOYD M. BYERS

Animal Science
Texas A&M University
College Stn., TX 77843

LYDIA ZEPEDA

Agricultural Economics
University of Wisconsin
Madison, WI 53706

NABC/JOYCE FELLOW

CASSANDRA KLOTZ

Agricultural Economics
UC Davis, Davis CA 95616

ISSUE 1) We need to assess the compatibility between broader social, value, political, economic, cultural and spiritual dimensions and the technology.

ISSUE 2) We need to assess the *process* by which animal, human and environmental safety issues of AGPs are evaluated in order to possibly improve that process.

ISSUE 3) We need to provide access to the information on the risks and benefits to the groups affected by the specific AGP.

ISSUE 4) We need to eliminate international regulatory disparities with a goal of free/fair trade.

THE BROADER SOCIAL ISSUES

The first goal was the most controversial in the sense that it was the most difficult to reach a consensus on. There was strong disagreement about introducing broader social issues as a "fourth hurdle" to technology assessment. However, *agreement that some assessment of the broader social impact should take place at some point* arose from the following issue developed by a workshop subgroup: "What are the social value, political, economic, cultural and spiritual systems that are the context for new technologies; recognizing that 1) animal growth promotants are not unique and 2) all technologies are expressions of the system in which they are developed?"

This issue reflects the observation that technology is a product of the culture, the values of its developers and/or the society from which it evolves. The cultural context and values in which technology is developed is often not acknowledged as influencing that technology. Hence, technology is not value-free. This created conflicts between groups affected by the technology who may have different values. Thus there is a need to recognize values in technology and its applications, and when other values might be inconsistent with that technology. There was some contention whether or not major incongruences exist between social value systems and technologies, and whether it is possible to agree on a definition of "compatibility."

An example of actual conflicts is development of pork products in a Moslem country. Other conflicts of values may not be so obvious. Thus, a consensus was obtained for the goal of "identifying and measuring these broader social factors," and that the best way to accomplish this was via research.

Disagreement occurred as to who should do the research, whether public or private funds should pay for it, and how it should be applied. Within the subgroups, it was felt public funds would be most appropriate because of the need for a balanced public approach. When discussed by the entire workshop, however, some participants felt private funds may be appropriate in some cases.

A recommendation by the subgroups to incorporate these social impact assessments into technology evaluation and institutional decision making met with a great deal of controversy. Generally, agribusiness industry representatives, producer groups and university natural scientists felt socio–economic assessment would prevent research and/or product development if it were applied as a criterion. One participant was concerned that socio-economic assessment could mean "social critics are free to operate without restraint and do not have to measure up to any standard," rather than peer reviewed socio–economic research. There was also disagreement as to when such analysis ought to occur and who ought to do it. It was pointed out that it may not be feasible to do research on impact assessment before product development, or prior to product release. Particular concern was raised in the discussion over whether to analyze these issues prior to or after approval of the products, and also pre– or post–product development.

Some social impact assessments are currently being done, however, in an ad hoc manner. One participant pointed out that farmers make an ad hoc assessment through their purchasing decisions, which reflect a variety of social values. Concerns and questions were raised about whether current assessments by researchers are being done appropriately, and how they should be done. How and when the information from the assessment should be disseminated was also a concern, as was who should do it and how it should be funded.

EVALUATION OF THE REGULATION PROCESS

Safety and regulation issues were combined to form an issue: "*The process by which animal growth promotants are regulated for human, animal and environmental safety should be assessed for possible improvement.*" The issues were originally expressed separately; "assurance of product safety" and "is the regulatory process to evaluate AGPs applied appropriately, correctly and legally?" However it was determined one could not do one without the other, so the issues were combined.

There was no consensus about what ought to be improved or how it might occur. However, concerns included: credibility of decisions made by regulators; the definition of safety levels; the timeliness of the process; the openness of communication; the timing of public input; and the disclosure of information versus proprietary rights. The subgroup premise to this issue is that two major concerns of consumers are price and safety.

The methodology proposed to accomplish this goal is to a) identify the concerns of the different constituencies, possibly in a meeting or via a survey; b) brief the groups on the regulatory process as they each may not have a complete overview of the process; c) develop a white paper that reflects the concerns of the groups with subsequent feedback in order to reach a consensus; d) develop solutions for the identified concerns.

Seven groups of constituents were identified: regulators, consumers; industry; producers, academia, researchers, educators; and legislators and environmentalists. One critique of this process is that it might take too long. Another was the difficulty of handling a lack of consensus and the level and timing of communication among the constituents.

There was a lack of consensus in the workshop about the wording of the issue: "should be assessed for possible improvement" versus "could be improved." Some felt more comfortable with the latter because they thought improvement was needed. The former working was adopted, however, indicating a lack of consensus about the need for regulatory improvement.

Other concerns arising from this issue were whether the Food and Drug Administration is funded well enough to do its job. Some participants had questions about how technologies and products should be regulated, how to make the process more credible, and how might the good aspects of the regulatory process be communicated to the public to improve credibility of the regulatory agencies.

INFORMATION ON THE DISTRIBUTION OF RISKS AND BENEFITS

Consensus was easily reached for another issue, *"Access to the information on the risks and benefits should be available to the groups affected by AGPs."* However, there was no consensus on who should acquire the information, when it should be done, and whether it should affect regulatory restrictions.

The methodology proposed involved research, education and public policy. "Credible" research would identify the specific groups affected to determine potential risks and benefits. A public education program would disseminate the information and public policy would be set to insure that relevant, credible, objective, nonproprietary information would be available to all.

The word "objective" was a point of contention (as it was in Issue 4). It lead back to Issue 1, in which "objective" information depends upon the values of the individuals or institutions supplying and receiving the information. Along a similar vein, concern was expressed over individual opinions and values dominating risk and benefit standards. Some felt there is a need to develop a *systematic* way of obtaining and disseminating information of risks and benefits. One person expressed concern over information "carpet–bombing," i.e., the ability of groups or individuals to dominate information dissemination about animal growth promotants, presumably through the media.

INTERNATIONAL REGULATORY DISPARITIES

This issue stemmed from concern about who is making decisions for whom. The issue is *"International regulatory disparities affect trade and developing country economies."*

The methodology proposed to accomplish the goal of eliminating disparities is: a) establish an international dispute settlement mechanisms such as GATT; b) conduct research in a neutral, independent setting such as international research centers; c) establish objective international standards such as CODEX; and d) educate consumers possibly using UN/FAO funds.

Disagreement occurred over the goal of "free" trade versus "fair" trade. Some felt "free" trade would benefit countries and industries with large market shares that are able to undercut smaller industries and countries through "unfair" trade practices. In particular, the potential of large, developed countries to overwhelm smaller, less–developed economies was a concern of some. One person stated that free trade is means to a goal and not an end in itself.

The methodology proposed by the subgroup raised some controversy as to who ought to pay; whether any international body is really "neutral;" and whether objectivity of standards is possible given the many assumptions and models required to make assessments. In particular, the neutrality of CODEX was questioned. Questions were also raised about how to establish neutrality and to develop objective standards.

SUMMARY

Many of the situations, concerns and issues emerging from the animal growth promotants workshop transcend the topic of AGPs and even biotechnology, and deal with the broader issue of technology appraisal. There was a consensus that the issues are rarely product specific and that future workshops might be aligned to reflect this.

General concerns voiced by the workshop include the uncertainty of the new technologies, the availability, source and communication of information about AGPs, and trust or credibility in institutions, the regulatory process and the media. These themes of uncertainty, information and credibility emerged in many of the issues, from regulation to labeling.

Another recurring theme was perceptions and values. It was recognized that different constituents have different perceptions and values. However, value judgements were invariably placed on whether these perceptions or values are "good" or "bad." As an example, many scientists perceived science as "objective." However, it was pointed out that science is a product of the assumptions made, choice of experimental design, choice of research model, and choice of statistical analysis used, all of which are subjective decisions.

The conflict between perceptions and values was most evident in participants assumptions about the role of agriculture, and hence agricultural policy and research in our society. Production maximization, profit maximization, sustainable agriculture, and providing top quality food and fiber were all evoked as goals of agriculture. Some viewed production maximization and profit maximization as equivalent, however this is incorrect, unless input prices are zero. Profit maximization and sustainable agriculture were seen by some as divergent goals in agriculture, despite the fact that unsustainable practices are not profitable in the long–run and agriculture is not sustainable unless is it profitable.

Agriculture is varied and involves many people and institutions with different cultures and values. Different approaches are likely to exist and to be encouraged. However, without some agreement about the role of agriculture, it is unlikely that a consensus can be reached on substantive issues of agricultural policy and public funding of agricultural research with respect to AGPs. In particular, some interested groups will continue to feel disenfranchised, expressed by an overriding theme of "who is making decisions for whom?"

The animal growth promotants workshop had representatives from universities from a variety of disciplines, from agribusiness industries, producer groups, environmental groups and consumer groups. Two groups conspicuous by their absence were regulators and individual producers.

VISIONARY PATHWAYS

c. 1968

THE PAST AND FUTURE OF AGRICULTURAL BIOTECHNOLOGY

Some time ago, at about the same time that I was invited to speak at this meeting, I read an article about the invention of agriculture. It was a report on the work of a team of archaeologists at Yale, who believe they have pinpointed the origin of agriculture in time and space: about 10,000 years ago, at the north end of the Dead Sea. This area had already been identified as the location of an unusually advanced civilization. People had been gathering grain there for a long time. But then they made the transition—and this must be one of the great transitions in the entire course of human evolution—to planting grain and cultivating it.

WALTER TRUETT ANDERSON
1112 Curtis Street
Albany, CA 94706

Political scientist, environmentalist, journalist and author, Anderson has been active in environmental politics for more than 20 years. Currently environmental editor of the Pacific News Service, he is also a Fellow & faculty member at the Western Behavioral Sciences Inst. in La Jolla CA.

The interesting part of the story is why and how they did this. According to the Yale archaeologists, the civilization—which they call the Nataufian—was an unusually stable one for the time. The people were no longer merely nomadic hunters and gatherers. They had well–built houses and a sophisticated social structure. They knew how to harvest wild wheat and barley with flint sickles, and they processed the grains with stone mortars and pestles. They supplemented their grain diet with meat from deer and other wild animals.

Then the climate changed, as climates do. Summers in the Jordan Valley region became hotter and more arid. Some of the water sources dried up, and people began to congregate around the larger lakes. Summer droughts damaged game habitat and shortened the growing season. So there were food shortages, and migrations of displaced people; all the classic symptoms of a civilization on the verge of collapse.

But this civilization did not collapse. The crisis also presented an opportunity. Although the transformed climate was tough on the perennial plants, it favored the annual species of wild grains and legumes—the ones that completed their life cycle in the late spring, left big seeds that could survive the summer drought and germinate at the beginning of the next winter growing season.

Agriculture was probably invented many times, by different people and in different places.

Probably the Nataufians already understood the mechanics of this process, and it was a logical step to begin to help it along—make it more dependable and productive—by saving seeds in the summer and planting them at the start of the wet season. And so began the ancient rhythms of agriculture, with times to sow and times to reap.

Agriculture was probably invented many times, by different people and in different places. But I found this article fascinating and revealing in many ways; it tells us something about where we all came from and if we look closely, it also can tell us something about where we are going.

The beginning of plant sowing and harvesting was also the beginning of plant breeding. The archaeologists believe it was easier for people to harvest certain grain plants—mutants with larger seeds, with tougher connections between seed and stalk that made the seeds less likely to scatter when people gathered the stalks and carried them to storage areas. In as few as 22 years, according to one hypothesis, the cultivated fields may have been taken over by the mutant varieties. So when the course of evolution changed for the human species, it changed for other species as well.

Agriculture then spread northward into Turkey, and eventually into Mesopotamia. As it did, it led to ongoing waves of change. Landraces of cultivated crops evolved. People cleared out weeds and altered animal habitat. Eventually they invented irrigation, which launched another round of effects on soil and water systems—and probably on the local climate.

About a thousand years later—at least according to the source I am citing here—people started domesticating animals. And of course that brought its own series of environmental and evolutionary consequences as people got into selective breeding, and battled predators, and altered ecosystems. Some of the ecosystem changes were deliberate—such as when people cleared trees to create savannas for grazing animals. Some were accidental—such as what resulted when people began to move herds of animals outside of their original habitats. If you have ever watched a grazing herd of sheep or goats, you know what I mean about ecological impacts.

human beings have a superior ability to adapt to a variety of ecosystems

I am getting away from the Nataufians here, and I will return to them in a minute. But first, let us pull back the camera and take a quick look at what is generally known about the links between the development of human civilization and impacts on what we might call nature. We see homo sapiens emerging as a distinct species and migrating all around the globe. Why? Because human beings have a superior ability to adapt to a variety of ecosystems. How do we adapt? We adapt by inventing new techniques of survival, such as agriculture and we adapt by modifying ecosystems to suit our needs. Every living thing, even the humblest lichen on a rock, modifies its environment. But no species comes anywhere close to homo sapiens in terms of environmental impacts. We are, as Isaac Asimov once put it, the environmental modifiers par excellence.

Human beings moved about the globe and moved plants and animals about the globe. Rice, corn and wheat became world citizens, and so did Holsteins and Herefords. The world changed.

A lot of those changes took place in prehistoric times and we do not know that much about them. The transformation of North America took place more recently, and it is very well documented. I started poking into this subject some years ago when I was writing a textbook on American government and discovered in my researches something that I had not understood before about our history—something that fascinated me so much I later wrote a book about it. The discovery, simply put, was that the American colonists not only established a new society and a new government here, but in the process rebuilt the continent into something more suitable to their purposes. They imported plants and animals. They battled native weeds and predators. They cleared forests, They changed waterways. They dredged harbors. They built canals. They laid out roads. The single item that intrigued me the most was that the Mayflower had not only human pilgrims aboard, but also pigs and sheep and cattle—and a few stowaways, like dandelions in the food larders and moths in the woolen clothes. So there are pigs, weeds and bugs whose ancestors came over on the Mayflower. And practically everybody contributed to this rebuilding effort in some way or another. It expressed a great social consensus about what needed to be done.

And when people went west, the wave of change rolled across the continent with them. We have American myths to commemorate Paul Bunyan clearing the native forests and Johnny Appleseed planting immigrant trees.

Here in California the ecological transformation proceeded with tremendous speed after the Gold Rush. Undoubtedly the Indians had made their own imprint before that—do not believe that stuff about Indians having no impact on nature—and so had the Spanish explorers. They accidentally brought in grass seeds that established themselves well here, and they also brought, as a food supply, snails whose descendants are at this very moment chewing at the

American colonists not only established a new society and a new government here, b in the proces built the con nent into something more suitabl to their purposes.

We are living in the midst of the fastest and farthest reaching ecological change in human history.

vegetables in my backyard. Thousands of people came to California. Miners dug in the mountains and washed away tons of soil with hydraulic mining devices. They hunted game and put some native species, such as the California grizzly, on the road to extinction. People built towns and cities, cutting trees to build them. Herds of cattle and sheep were brought in to feed the fast-growing new population. So it went, and so it goes.

Now, the things I have mentioned here are just a few examples of what has been going on as long as there have been human beings on the earth. Wherever human beings live, the environment changes. As human populations grow and their technologies become more advanced, the changes happen more quickly.

We are living now in the midst of the fastest and farthest-reaching ecological change in human history. Populations growing and expanding into areas that used to be wilderness. Plants and animals moving about the planet. Species extinction and genetic erosion proceeding at exponential rates. Tropical forests being cut down. Aquifers being depleted. Pollutants being dumped into the rivers and the oceans.

You have no doubt noted that I started out talking about agriculture and am now talking about the environment. When did the subject change? It didn't.

Agriculture, civilization, environmental change and environmental problems are all part of the same process. Call it human evolution. For a long time, people developed agriculture and civilization without paying much attention to the environment part. The environment simply did not exist as a concept for ancient peoples—even people of a few decades ago—in the way it does for us today. Eighteenth-century political philosophers such as Montesquieu talked a lot about "climate," which meant roughly the same thing as "environment" but what they were mainly interested in was the question of how environment affected civilization, what kind of an environment was most likely to become the basis for an advanced society. They did not have much to say about the civilization affected environment. The news had not yet begun to come in.

In the middle of the last century, a great and original American scholar named George Perkins Marsh wrote a book entitled *Man and Nature,* or, *Physical Geography as Modified by Human Action.* It summarized a great number of separate papers and reports that had been written up to that time—Marsh read twenty languages—about the impacts of ordinary human actions such as farming, logging, fire, converting land to agriculture, modifying waterways, domesticating plants and animals. It was a blockbuster of a book. It made a profound impression on many people, both in the United States and Europe, and in a way it is still making an impression. Marsh was not a back–to–nature man, he did not say that people should go back to being hunters and gatherers. He merely reported on the impacts—but because most people had never thought systematically about those impacts before, the news it brought was most unsettling. Marsh is generally regarded as the source and inspiration of the early conservation movement that eventually came to be associated with such people as Theodore Roosevelt and John Muir, and it would not be stretching things at all to call him the grandfather of the modern environmental movement.

Nowadays just about everybody is an environmentalist...

Nowadays just about everybody is an environmentalist, and everybody has his or her idea of what environmentalism is about. Some people say it is about preserving wilderness. Some people say it is about conversing resources. Some say it is about spirituality. Some say it is about Gaia. I submit that what environmentalism is basically about is feedback. It is about information. It is about the discovery, a fairly recent one, that what we call by various names such as human cultural evolution, civilization, or progress have impacts on nature. That is what gave rise to environmentalism to begin with, and that is what is going to produce a lot more of it, because both impacts and feedback are on a strong growth curve.

A lot is being written and said these days about the impacts, and I will merely summarize briefly. First, human population is growing at a rate that would have been utterly unthinkable to most people in Malthus' time—but probably not to Malthus. We add well over eighty million people to the world's total population every year. Every year—to put it another way— the world faces a situation that is significantly different, by over eighty million people, from any situation it has ever faced before. Many of the major environmental problems that I mentioned above are getting worse very rapidly. We are making some headway in a few areas, such as cleaning up some American waterways, but unfortunately most of the major environmental problems in the world are quite out of control.

Secondly, at the same time that our human numbers and our impacts on the environment are increasing, our science and technology of environmental information–gathering and information–processing is improving tremendously. People are putting together new information–gathering organizations such as the International Geosphere–Biosphere Program, using new technologies such as satellite monitoring. In many ways, getting much, much better at getting feedback. This, too, can be expected to continue to grow and develop and

...the environment is almost entirely dependent for its importance as a political issue on abstract information.

change the ways we act and think.

It is worth noting that the environment is almost entirely dependent for its importance as a political issue on abstract information. If you stop to think about it, you will notice immediately that most of what we worry about and argue about in relation to environmental deterioration has to do with data and interpretations of data, news reports, projections and scenarios of the future.

At the top of the list of these environment–information worries is the greenhouse effect. It may well be the most important crisis the human species has ever had to deal with, but where is it? You can not walk out your door and feel it or see it. All our concern is based on projections and scenarios. The late Professor Walter Orr Roberts of the University of Colorado, who was my main source of information when I started looking into the matter, summed it up concisely. He said the buildup of CO_2 and other greenhouse gases in the atmosphere is scientific consensus; there is not much controversy about it. But there is no such consensus about whether the buildup will lead to a global climate change or what that change will be, or whether it has begun. Roberts said that the 1980s were the hottest decade since the invention of the thermometer. He also said that the difference was within the range of what could be called normal fluctuation. We do not know.

Roberts was one of the first scientists who made a specialty of looking into the possibility of global climate change, and he thought we were in the early stages of a warming trend caused by the greenhouse effect. So you might class him among the pessimists, but he was fairly optimistic about our prospects for coping with it. He never tired of pointing out how adaptable human beings are.

It seems to me that if you put all the pieces together—not only the indicators of global warming but also the data and projections we have on such matters as population growth, species extinction, soil erosion, aquifer depletion and deforestation—it adds up to a strong reason for believing that global conditions are going to be quite different in the years and decades just ahead of us, even if we can not say precisely how they will be different.

Up to this point I have said nothing at all about biotechnology. But as we all know, it is moving rapidly also. Perhaps not as rapidly as some people expected, but there is no doubt that we are looking at a full–scale scientific and technological revolution. I am on record as having said and written that the Biological Revolution will turn out to be as momentous as the Industrial Revolution, and I have seen nothing to make me change my mind about that. And I do not see any likelihood at all that that revolution will unfold separately from the large-scale global environmental and developmental concerns I have been summarizing here. We are entering a period of global crisis, and we are going to find it necessary to mobilize all of our resources—definitely including our scientific and technical know-how—to deal with it.

With that somewhat sobering thought in mind, let us go back to the Nataufians. The article I mentioned described their innovation as the result of a "convergence of accidents." Four seemingly–separate elements happened to be present at the same time—genetic resources, technology,
social organization and need.

The genetic resources were available wild grains. The technology was the Nataufian's knowledge of how to harvest and process grain. The social organization was essential. Had they been a more primitive nomadic society, they probably would not have been able to organize the labor and the distribution of food. And need. They were in a time of crisis and they knew they had to do something.

So they invented agriculture. Is it possible to compare our present situation to that of the Nataufians? Yes. It is always dangerous to make historcal connections of this sort, especially over a span of ten thousand years, but it is even more dangerous to fail to learn from history.

We can construct a plausible scenario of the near–term future that is based on a convergence of the same four elements. Genetic resources, technology, social organization and need. Genetic resources that can be adapted to new uses. Technology—biotechnology—that makes it possible to do that. A social structure that supports research and development. And need—a crisis situation, such as global climate change combined with overpopulation and other environmental stresses, that makes improvisation necessary. A situation roughly comparable to that of the Nataufians. This time, the response is not the invention of agriculture, but a great range of more effective ways to utilize biomass to produce food, fiber and energy.

Some such future situation is quite likely to unfold. But we should note that each of the four necessary elements may not be up to the occasion. The loss of genetic resources, as you all know, has become a serious international issue in recent years. Our technology and social organization may not be ready. Even the matter of need is in some ways problematic.

The Nataufians were dealing with a situation that was already present. The

...we know about environmental impacts. We do not know how to predict them well...

need was visible and they saw it around them. If the scenario that I am suggesting here is to become a reality, in any way that makes it a success story and not a global tragedy of a scale we have never before seen, it will have to be based on a response to anticipated, rather than present need. The consequences of not being ready, not being able to respond to the situation in time, are likely to be disastrous. When Walt Roberts and I were discussing this, we talked about the fact that, up till now, human adaptability has been a matter of dealing with environmental conditions that already existed. The human species has so far has not shown much ability to anticipate.

So there is serious doubt as to whether the "social organization" part of the scenario is really up to the challenge. Our political system is not famous for long–range planning. Our economic institutions are mainly keyed to need expressed in terms of market demand—either already existing or likely to exist soon enough to justify capital investment.

Another thing that makes us different from the Nataufians is that we know about environmental impacts. We do not know how to predict them too well, but we have every reason to believe that any large–scale transformation of agriculture such as the one I just suggested would also have many secondary and tertiary environmental impacts. And this concern about environmental impacts has a lot to do with how we make progress in science and technology.

Biotechnology is unique in having evolved within the environmental era.

Biotechnology is unique in having evolved within the environmental era. The work that produced the first successful recombinant DNA took place in the early 1970s, and the public started hearing about it a year or two after the first Earth Day and the United Nations Stockholm conference—events that mark the beginning of the modern environmental movement. Much of what environmentalists have to deal with involves the results of technology—cleaning up the effects of the Industrial Revolution that people either did not know about or swept under the rug—and so it is hardly surprising that there is a strong anti–technological bias to the environmental movement.

I doubt that environmentalist concerns about biotechnology are likely to diminish or disappear. They will be around for several reasons. One reason is that some people have made a career out of frightening people about biotechnology and would be lost without it. Another is that some people have made a religion out of being frightened and are no more interested in letting go of their scenarios of biotechnology on the rampage than right-wing true believers are interested in letting go of their scenarios of global communism on the march.

And a third reason is that there are excellent scientific reasons for being concerned about adverse ecological impacts from genetically modified plants or animals or microorganisms. I expect and hope that organizations represented here, such as the Environmental Defense Fund and the National Wildlife Federation and the Sierra Club will continue to demand adequate risk assessment and regulation. So I do not think a get-out-of-the-way-and-let-us-save-the-world argument is going to prove to be very persuasive with environmentalists—especially if your idea of saving the world turns out to be developing herbicide–tolerant crops.

Technological fixes... will not only be tolerated, they will be demanded.

And on the other hand, a lot of people are more interested in biotechnology's ability to deliver than they are in the possible adverse consequences. We have seen, in the case of AIDS, tremendous pressure put on the scientific–industrial establishments to come through with a vaccine or a cure. Many of you are probably familiar with the book *And The Band Played On*, which is a powerful and angry indictment of the medical–research establishment for failing to move more vigorously against AIDS. If the kind of global environmental crisis that I have described begins to become inescapably obvious and not just a matter of academic debate, you can expect to hear a great clamor to mobilize science and technology to deal with it. Environmentalists are suspicious of technological fixes, but the general public has no such reservations. Technological fixes will do fine. They will not only be tolerated, they will be demanded.

This means that agricultural biotechnology will find itself in the exciting but uncomfortable position of being pushed and pulled in different directions at the same time—asked to solve the world's problems while simultaneously being accused of getting ready to cause perhaps worse problems.

Some years ago, in his book entitled *Broken Code: The Exploitation of DNA*, Marc Lappé said that biotechnology occupies a "special moral position." I think he was quite right. And it is not entirely pleasant to be in a special moral position. It means that people ask a lot of you, and hold you accountable in different ways.

You are in a special moral position, you who are in the process of creating agricultural biotechnology. Rather, I should say we are in it—all of us who are in some way taking part in the dialog in whatever role, whether as scientist or industrialist or critic or regulator. If in the future somebody writes some book like *The Band Played On* about agricultural biotechnology's failure to anticipate and respond to the world's needs in a time of ecological and developmental crisis. I

You are in a special moral position...

do not think the anti-biotechnology crusaders are going to come out of it looking any better than the alleged mad scientists.

How to respond to this moral challenge? There are many things that need to be done—and many things that people are already doing. For environmentalists, it means dropping the "Great Satan" view of biotechnology and putting our risk–assessment and regulatory concerns in the context of a policy that actively supports research and development. The Ecological Society of America did this in its 1989 paper on the environmental use of genetically engineered organisms. The World Resources Institute did this in an excellent recent report entitled *Transforming Technology*, which explored, among other things, the compatibility between biotechnology and "alternative" or "sustainable" agriculture.

A lot of people in the environmental community have bought into a polarizing and, I think, foolishly oversimplified idea that the world is either going to move toward a future of "alternative" or "sustainable" agriculture that is easy on the environment and helpful to the disadvantaged, or toward a mechanized and expensive high–tech agriculture that is resource–wasteful, polluting and of benefit only to big-time agribusiness. Personally, I think we are going to find agriculture becoming as pluralistic as the rest of society, with a vast range of techniques and no clear either–or. The World Resources Institute report speaks to this when it describes high–tech agriculture that is resource–thrifty and ecologically sound. "Farmers of the next decade may grow genetically engineered disease–resistant soybeans on one conservation–tilled field while monitoring real–time soil moisture data from a nearby insect–resistant cornfield and analyzing feed requirements for their cattle by computer."

There are many ways that biotechnology can contribute to the needs of small farmers in economically disadvantaged areas.

There are many ways that biotechnology can contribute to the needs of small farmers in economically disadvantaged areas. But if that comes about it will be because people in agricultural science and industry direct more of your energies toward environmental concerns and Third World development—even toward projected needs—as well as toward the commercial possibilities in Western developed countries. This is the challenge to the other side, and there are many ways to meet it. *Pro bono* research, for example. New courses and research programs and even special institutes in the universities. More support for federal agencies such as the Agency for International Development and for international initiatives such as the U.N.'s International Centre for Genetic Engineering and Biotechnology, which the U.S. government has studiously and deliberately neglected.

There are a lot of ways that environmental and public interest groups and universities and biotechnology industries can work together—even while we continue to argue about the safety of bromoxynil or the impacts of genetically modified microorganisms—to advance the general human and ecological well-being we all claim to be concerned about.

There are a lot of ways that environmental and public interest groups and universities and biotechnology industries can work together.

Perhaps I am overstating the urgency of the whole situation, maybe it really is not—and is not about to become—as serious as I have indicated. So let me sketch out some alternative futures.

In closing, I will briefly describe three scenarios of change that I constructed for a conference of planning officials in Hawaii two years ago. You can pick the one that seems most plausible to you.

The first scenario I called "Business as Usual." In this history of the future, instabilities resulting from environmental disturbances, population growth and the needs of developing countries are minimal. Assumptions about economic growth, international development and human progress remain essentially the same as they were during the post–World War II era.

In the second scenario, entitled, "Disturbance and Adaptation," there are serious global environmental challenges, including warming as a result of the greenhouse effect and they are met by a range of reasonably successful responses—including responses based on advance planning. There are costs and changes, but we rise to the occasion.

The third scenario I call "Chaos and Conflict." It shows environmental changes too great or too rapid to be handled, with consequent breakdown of social and political order. One likely result would be new terrorist or revolutionary movements, as environmental concern escalates into environmental fanaticism.

Fill in these outlines with whatever information you have, and take your choice of which you believe is the most likely. Most of my remarks have been based on my own expectation that the future will be somewhere in the margins between the second and the third scenario, between the "Disturbance and Adaptation" future and the "Chaos and Conflict." In fact, as I see it, that is where we are already. I often quote as my favorite futurist the ex-Kansas City Royals reliever Dan Quisenberry, who once said, "The future is just like the present, only longer."

Now I realize that there are all kinds of pressures and demands on all of you—what we might call real-world considerations—that make it impossible for you to be completely preoccupied with the sort of concerns I have been talking about here. I understand that, and I am not here to ask you to quit your jobs or drop your present research programs, or ignore the interests of your stockholders. All I want to say is that the world I have been talking about—the one with 5.4 billion people on it, with eroding soil and depleted aquifers and disappearing forests and polluted water and a climate that may well change significantly in our lifetime—that is the real world and it can neither be ignored nor separated from all of the other things we do.

So I suggest that we look for ways to show we know that and to find ways to make that awareness a visible and integral part of our work—whatever our part in the Biological Revolution may be. And I think that if we do, we may be able to cut through some of the conflict that has held us back and that biotechnology will progress and flourish and be seen in the future as one of humanity's great accomplishments—as I sincerely hope it will.

BIOTECHNOLOGY AND THE ENVIRONMENTAL VISION

In the spirit of Walter Truett Andersen's talk, I hope to provide a responsible contribution to the ongoing debate whose long evolution he has described.

I have enjoyed the opportunity occasioned by this speech to take a somewhat longer view of biotechnology; to reflect on how far it has come and where it might be going. In so doing, I have pondered two questions. The first is the question of my vision of the future of biotechnology. The second concerns the idea of a crossroads—whether, indeed, biotechnology is at a crossroads.

MARGARET G. MELLON

Director, National Biotechnology Policy Center
National Wildlife Federation
1400 16th Street, NW
Washington, DC 20036

Mellon, Ph.D., J.D., evaluates and monitors biotechnology from an environmental perspective. She often testifies for Congress and frequently speaks on issues in biotechnology.

On the question of my vision of the future of biotechnology, I need at the outset to rephrase the question, for I do not have a vision of the future of biotechnology. Biotechnology is a collection of tools, capabilities and products. In my view the vision should focus not on tools, but on how we want to use them.

My vision is not of biotechnology but of a new kind of agriculture—a sustainable agriculture—with characteristics and objectives different from the industrial agriculture currently practiced. The important question is whether the tools and products of biotechnology advance us towards that vision.

What is the vision? It centers on prosperous farmers, an abundance of safe food and a clean environment. In it, clean—not polluted—water washes off agricultural fields and percolates through the soil to the groundwater below. Wildlife thrives at the edge of fields, rather than dying from consuming pesticide pellets. Weeds are in the fields, but kept to levels that do not interfere with yields. Cultural practices and intelligent management control the interaction between pests and crops. Agriculture de-

Biotechnology is a collection of tools, capabilities and products

velops in response not to short term goals—tempting as that always is—but with an eye to long term productivity and prosperity.

Such an agriculture will depend on methods that will be as viable 50 years from now as they are today. It will look for ways to use and reuse materials on the farm. It will put a premium on information and management rather than on purchased inputs.

Such an agriculture will be supported by a revitalized and redirected land grant university system. The new land grant universities will be able to provide information about cultural and integrated farming practices as readily as today's land grant universities disperse information about chemical pesticides. These new land grant universities will command a much larger part of our national research pie because they will be seen as ensuring not only agricultural productivity but a healthful environment, safe food, and a connection to nature for our increasingly urbanized populations. Land grants with this new broader vision of their mission would be at the forefront of our national life rather than in the backwater.

My vision for the future of agriculture also includes a varied, abundant and safe food supply with which people are comfortable. Consumers would not need to wonder whether the carrots they picked up at the market had mouse or human genes in them. Nor would they need to worry about whether the tomatoes they were eating contained a toxin gene inadvertently transferred into the tomatoes by careless genetic engineers.

Where does biotechnology fit into this picture? So far, at least, biotechnology will not substantially advance us toward a sustainable agriculture. Let's look at the four categories of products that are the subjects of workshops at this meeting. Herbicide–tolerant crops, animal growth promotants, transgenic animals and biological control agents. Three of the four—herbicide–tolerant crops, animal growth promotants, and transgenic animals—could easily be dispensed with without retarding our progress toward a sustainable agriculture. The fourth, biological control agents, is an important component of a sustainable agriculture but biotechnology will have only a minor impact on the efficacy and adoption of such agents.

My vision for the future of agriculture also includes a varied, abundant and safe food supply with which people are comfortable.

I am not saying that the products of biotechnology have no place in a sustainable agriculture. Some, for example pest–resistant crops, could be important components of such systems. But right now, in 1991, the available biotechnology products are not contributing significantly to the development of sustainable agriculture. Some, such as herbicide–tolerant crops, are antithetical to that vision.

Biotechnology is dampening our progress towards sustainable agriculture in a more subtle way, meaning that biotechnology is popular with government, universities and private industry because of its potential to generate new products. For example, companies and universities currently spend millions to develop new biological pesticides. In general, these products—although not without problems—are environmentally preferable to existing chemical pesticides. But at the crux of the matter, the new pesticide products are not the best solution to the problem of pests. It is widely agreed that systems approaches—for example, crop rotation and other methods—could avoid the need for the majority of pesticides, both chemical and biological, now and into the future. The problem is that the hype surrounding biotechnology diverts our attention from those solutions by focusing attention on technologically dazzling new products. The bias towards products deprives the systems–based approaches of the research and extension resources that are required to achieve their full potential.

It is vital that agricultural policy put biotechnology in a position to serve and not displace sustainable agriculture.

It is vital that agricultural policy put biotechnology in a position to serve and not displace sustainable agriculture. The question should not be whether biopesticides are compatible with sustainable agriculture, but whether, after crop rotations have been employed to their fullest, which pesticides are still needed.

To offer a metaphor for nature of the choice we have before us—biotechnology can be considered an elephant standing in front of two tents. One tent is large enough for the elephant, but it is also somewhat rickety. It has stood for a while, but will not last much longer. It is the tent of industrial agriculture.

The other tent is smaller, not as glitzy, but sturdy. It will last as far into the future as we can see. But it will not hold the biotechnology elephant. If the elephant charges in, the sustainable agriculture tent will come tumbling down. What is needed is to put the biotechnology elephant on a diet, to cut it down to size so that it will fit into the sturdy tent. Specifically, the

Despite almost a decade of federal level effort, the biotechnology regulatory system is in a shambles.

United States Department of Agriculture (USDA) should not be distributing buttons emblazoned with "Biotechnology is the Future of Agriculture." Biotechnology cannot be the future of agriculture. Technologies have no values; they embody no goals. And yet they are not neutral in their application. Technology can influence outcomes, particularly if the tools themselves are mistakenly regarded as goals.

It is time to articulate a vision of sustainable agriculture and ask how best to achieve it. Our goal should not be merely more new pesticides but a 75 percent reduction in pesticide use. By setting the proper goal, we will avoid the danger of spending millions trying to genetically engineer ten "better" pesticides, when for far less we could have taken our agriculture systems off the pesticide treadmill forever.

My second point concerns the idea of a crossroads. Is agricultural biotechnology, specifically the industry sector concentrated on transgenic plants and animals—really at a crossroads? In some ways, I would say not. The molecular biological and genetic sciences underpinning the technology are advancing at an accelerating pace. New scientific discoveries are leading to a stream of new products. The pace is slower than promised by the optimistic early entrepreneurs—but as the numerous field tests demonstrate, new products, especially engineered crops, are coming. A meandering path, perhaps, but not a crossroads.

A crisis looms for agricultural biotechnology in the area of product commercialization. This crisis could constitute a genuine crossroads. If the products now in the pipeline can make their way soon to the marketplace, the industry will develop confidence and attract new investors. On the other hand, if the early products are delayed in getting to the marketplace or are received unfavorably, the whole industry would be set back. The key is the federal review system. Without credible government review to assure safety, the pioneer products of biotechnology will never make it to the marketplace.

Despite almost a decade of federal level effort, the biotechnology regulatory system is in shambles. With regard to animals, such as fish, regulation is non-existent. With regard to food, authority exists and the government is aware that policy is needed, but it is proceeding at a glacial pace. Even where it appears to be functioning, the government is sometimes at the edge of its capabilities. In the crop plant area, for example, the USDA has overseen more than 200 field tests of engineered crops under the Plant Pest Act. While those tests generally have been conducted efficiently and safely, the Agency is facing an impasse in providing approvals for commercialization. The impasse is primarily attributable to limitations in the Plant Pest Act. Whatever the reason, the USDA has failed to describe a program for commercialization approvals. Open issues are legal authority for regulation, data requirements and opportunities for public comment. Lacking risk assessment protocols for large and commercial scale release, it is not surprising that the small scale tests done so far have not yielded many data on environmental risks. New rounds of small scale tests will probably have to be done to generate those data.

Our vision should focus on a low–input sustainable agriculture, not the course of a particular technology.

This is just one of the uncertainties facing those who want to commercialize transgenic crops. For example, later this month Environmental Protection Agency (EPA), USDA and the Food and Drug Administration (FDA) will meet for the first time to discuss the potential statutory overlaps on herbicide–tolerant plants. If the agencies are just beginning to address these issues, it means that proposed and final rules are years away, at best. This regulatory quagmire is beginning to assume crisis proportions. The issue of when, if ever, a reliable regulatory road map will be ready is becoming the rate-limiting factor in product development.

I am of mixed mind about the slowdown in the rate of the technology's development. If it were to result in a shift of resources to the system approach of a sustainable agriculture, I would consider it a boon. But I doubt that would be the outcome. More likely scientists will become impatient and like Gary Stroebel, will go ahead with releases on their own. That prospect is one of great concern.

To finish, let me summarize my two points. First, the agricultural biotechnology industry appears to be approaching a genuine crossroads in the area of commercialization. If the government cannot come up with protective, credible regulatory programs soon, transgenic products will not reach the marketplace and agricultural biotechnology will suffer a major setback. Second, our vision should focus on a low-input sustainable agriculture, not the course of a particular technology. I hope that under the leadership of the land-grant universities, biotechnology will be made to play second fiddle in the orchestra of sustainable agriculture. If it does, we can look forward to a symphony of environmentally and practically sound agricultural practices.

BIOTECHNOLOGY AT THE FOREFRONT OF AGRICULTURE

Why do we look at biotechnology and why will it be at the forefront of agriculture? Quite simply, all of us are directly involved with agriculture in one way or another—we all have to eat. Given that the world's population will double in the next 35 years—from 5 billion to 10 billion people, we need to be concerned about how the world's food needs will be met. In caloric terms, we will have to produce more calories in the next 35 years than we have produced since farming began 10,000 years ago. Assuming a population growing in a geometric progression and that food production is growing in an arithmetic progression, Malthus predicted that we would soon outstrip our ability to feed a growing population. Fortunately, he has been wrong for a few hundred years.

JERRY CAULDER

President, Chair & CEO
Mycogen Corporation
5451 Oberlin Drive
San Diego, CA 92121

Chair of the Industrial Biotechnology Association, Caulder has chaired U.S. Office of Technology Assessment workshops and served as a member of expert panels.

It's not what we don't know that hurts us, it's what we know that ain't true that's gonna kill us in the end.
–Jerry Caulder paraphrasing Will Rogers

Why? Because the American farmer has adopted science and technology as rapidly as it has become available, allowing farm production to outpace population growth. Advances in agricultural technology can be categorized into five distinct areas:

- –Domestication of animals and plants
- –Mechanization
- –Hybridization or genetics
- –Chemicals — pesticides and fertilizers
- –Biotechnology

Biotechnology fits well with the previous four and will quickly be adopted. Interestingly, each stage of growth listed above has resulted in a higher standard of living, longer life expectancy and better and cheaper food. Our existence is now dependent on fewer than 20 species of plants—we must use all available resources to assure that species are genetically fit to survive under the wide range of environmental extremes. These 20 species must be protected from pests to assure that they can reach their full genetic potential.

Many people... have no idea where food comes from.

Biotechnology is central to reaching that goal. It is debatable whether we

...each stage of growth ... has resulted in a higher standard of living, longer life expectancy and better and cheaper food.

should be dependent upon so few plants, but we are. One species alone, rice, furnishes about 60 percent of the energy for half of the people on earth. Stated another way, 30 percent of the human energy in the world comes from rice. From that standpoint, we are vulnerable. We must do everything possible to protect that genome in rice, because it supplies a lot of energy. Biotechnology offers many options for achieving that goal.

Seventy percent of our food comes from five crops—five crops that stand between us and starvation. Agriculture faces a simple challenge—to continue to adopt new technologies and to encourage quantum leaps in science (such as those in biotechnology), and use that technology to stay ahead of this curve. When oil reaches $40 per barrel, it is strange how people respond by funding research on alternatives to energy, yet we continue to cut basic research dollars for agriculture. Less than a 60 days supply of food in the world is considered a surplus. Contrast that with considering a 200 year supply of oil a strategic reserve. In constant dollars, agricultural research is being funded at the same level it was in the mid 1960s. In 1939, 80 percent of the federal research and development budget went to agricultural research. Today it is less than two percent.

We now have at our disposal this new science called biotechnology that will allow us to continue winning the race to produce food more efficiently, as long as this science is adequately funded.

To provide another example, does this quote sound familiar? "We have recently advanced our knowledge of genetics to a point where we can manipulate life in a way never intended by nature—we must proceed with utmost caution in the application of this new found knowledge." This quote is from 1906, from a critique of Luther Burbank's hybridization work.

The need for agricultural technology is viewed skeptically, for every trip to the grocery store finds it full of food. Many people, including my 20 year old son, think corn comes from the A&P—they have no idea where food comes from. Yet the scientific base on which this efficiency of production is based is very fragile and requires such an understanding in order to survive.

Efficiency is central to agriculture. We strive to maximize production while minimizing input. Low input sustainable agriculture is an acceptable goal only if output is not affected. We continue to tinker with the most efficient industry in the world—a system which allows 98 percent of the population to move from farms to pursue other interests, while the other two percent produces the food which allows us to do so. We are adept at telling that two percent how to run their business.

All of the five technology categories mentioned earlier encountered considerable resistance when introduced, so what must we do to assure that biotechnology is used to advance the efficiencies of U.S. agriculture and the American farmer?

Less than a 60 days supply of food in the world is considered a surplus. Contrast that with considering a 200 year supply of oil a strategic reserve.

First, we have to change our mental attitude about what we currently consider a surplus of food. This "surplus" should truly be considered a "strategic reserve." A few years ago, when we were in a drought, food prices rose five to eight percent. With the agricultural business base at around $400 billion, a five to eight percent increase in price far outstripped what it would have taken to store enough food to counterbalance the effects of the drought. The current levels of "surplus" are only inadequate "reserves."

Second, we have to increase our basic research in agriculture. A flat budget for over the last 20 years is a national shame. New dollars have to be made available to attract our finest scientists back to agriculture. Scientists, like the rest of us, tend to gravitate towards where they are appreciated most. Lack of funding in agricultural research has meant that NIH and other fields have successfully recruited the first round draft choices in medical and other research areas, rather than agriculture. We must become more competitive in order to be able to competitively bring top dollars back into agriculture. My father used to say "the actions that get rewarded get repeated." We have to start rewarding the right actions to get the right results.

Third, we must have a system of patent protection which protects intellectual property rights, and a regulatory system that allows these developing products to progress in an orderly manner to the marketplace, within a reasonable time frame at a reasonable cost. We will kill biotechnology quickly without a clear and timely regulatory process.

Last, we must continue to strive to strengthen the scientific literacy of our young people. We are a scientifically illiterate society. When a substantial portion of our population believes in pyramid and crystal power and the White House is run by an astrologist, we are in trouble. We must upgrade the scientific literacy of our young people. Vocational agriculture is not acceptable training and is doing more harm than anything else, because agriculture is a science, just like physics and chemistry.

The current levels of "surplus" are only inadequate "reserves."

The second area where policy affects biotechnology is in the regulatory arena. How do we promote science, its related technology and the freedom necessary to develop to its fullest potential for the common good, while at the same time

Biotechnology can make enormous contributions to basic research in agriculture, in medicine, industry and a host of other areas.

assuring a skeptical public that their safety is being looked after ? Almost every technology we have ever had faced the same crisis. We live in a society that is dominated by science and technology. Yet not even 5 percent of the population is equipped to understand scientific reasoning or the scientific method. This failure has led to a blurring of the distinction between science and technology. Science is the production of new knowledge, while technology is the application of knowledge to the production of some product. Most people do not distinguish between the two, creating many problems for new technologies.

Science, in and of itself, is more or less innocuous. As soon as it leads to technology, it is not necessarily innocuous. No society yet has learned how to accurately predict the consequences of new technologies. Herein lies the dilemma—there are inherent risks to the pursuit of new knowledge. The risks are greater, however, if we do not continue the pursuit. Biotechnology is the new knowledge that is with us now, and the question becomes a matter of risks balanced against benefits. Biotechnology is currently facing this challenge.

Biotechnology can make enormous contributions to basic research in agriculture, in medicine, industry and a host of other areas. Equally enormous, at least to a few, is a host of related environmental, regulatory, ethical and national security concerns. Scientists involved with recombinant DNA work have been aware since the inception of their craft of the potential for both good and bad. Scientists have taken a very strong initiative on their own in monitoring their research in what is generally regarded as a responsible and effective manner. The early concerns in biotechnology were with the safety of the experiments themselves—assuring that they were appropriately contained, and that even if organisms did happen to escape, that they could not survive and affect the environment. History has shown that this phase was handled very well. The creation of the P4 laboratory achieved total containment.

Commercial biotechnology is now poised to contribute dramatically to health and welfare worldwide. Yet the public's understanding of the underlying science and its grasp of biotechnology's capabilities and limitations has not kept pace with and undermines the industry's development. Currently, science, technology and its practitioners are both admired and mistrusted, and this does not bode well for public policy decisions.

Concerned about correcting the widespread misperceptions surrounding biotechnology, I will stress three fundamental facts:

1– Today's biotechnology is only one point in a long continuum of scientific inquiry reaching back to beer and breadmaking thousands of years ago. Selective breeding by early farmers was genetic engineering—we have long used biotechnology.

2– Biotechnological techniques are largely extensions of processes that occur naturally. Clones are not artifacts cooked up in a test tube—cloning abounds in nature as a means of asexual reproduction. Identical twins are clones. Moreover, evolution has not stopped. Mutation is an integral part of all life processes, and not an invention by scientists. What is new is our growing ability to simulate nature in ways that can offer enormous benefits.

3– Biotechnology does not describe a single procedure or process, but encompasses a diversity of means for using living matter to develop useful products. Recombinant DNA technology, the aspect most familiar to the public, is but one of the vehicles for product development.

Biotechnology does not describe a single procedure or process, but encompasses a diversity of means for using living matter to develop useful products.

We must have as a mission the demystification of biotechnology for the public at large. Speak to the Kiwanis Club, the Lions Club whenever and wherever you have the opportunity. Biotechnology, particularly genetic engineering, must be demystified to ensure that the public accepts it.

Without public acceptance, support and encouragement, the applications of genetic engineering may be regulated out of existence. The long regulatory cycles that we currently experience will crush smaller biotechnology companies, thus focusing development in the hands of a few large companies who have the funding and time to wait out the regulatory storm. It is crucial that scientists and informed lay people emphasize to the public that genetic engineering is as natural as plant breeding. Genetic engineering is basically a method to crossbreed or hybridize different organisms, to graft one or several genes from one organism to another. The plant and animal breeder has a limited number of useful traits—genes—with which to work into his crop seeds and animals. Genetic engineering offers a much larger menu of potentially useful traits and can enormously speed up the process of hybridization.

A corn plant with one gene changed by genetic engineering is still a corn plant, with one gene changed. It is not anything else, and to compare it to an alien introduction such as kudzu or the gypsy moth is scientifically absurd. If we are to banish pestilence and hunger from this planet, we must all be tireless opponents of divisive and unfounded rhetoric and proponents of scientifically based realism. We must demystify biotechnology and get people to understand the difference between facts and fantasy, between evidence and anecdote, between experiments and experiences.

On the Frontiers

c. 1980

BIOLOGICAL BREAKTHROUGHS & BOTTLENECKS

PLANT BIOTECHNOLOGY, PLANT BREEDING, POPULATION BIOLOGY AND GENETIC RESOURCES
PERSPECTIVES FROM A UNIVERSITY SCIENTIST

CALVIN O. QUALSET

Genetic Resources Conservation Program
U. California, Davis
Davis, CA 95616

A Professor of Agronomy and Director of the GRCP, Qualset specializes in plant genetics and is active in policy evaluation in biological conservation.

Plant biotechnology is developing as an applied disciplinary field and has gained high-level attention in public and private universities, the U.S. Department of Agriculture—Agricultural Research Service, and, of course, in the private sector. There is increasing interest in applying new approaches from biotechnology to international agricultural development. Plant biotechnology will be discussed here mainly from the university point of view, especially land grant universities; however, the views expressed here are personal ones not necessarily endorsed by my institution nor the National Agricultural Biotechnology Council. A dominant theme in this meeting is to assess the apparent lack of progress in delivering products of plant biotechnology to the marketplace for the benefit of consumers. It will not be my goal to enumerate the status of plant biotechnology accomplishments, although there have been many, but to point out some of the factors that impact upon scientific progress in plant biotechnology.

DEFINITIONS AND LONG-TERM COMMITMENTS TO PLANT BIOTECHNOLOGY

In assessing the progress of plant biotechnology it must be clear that there are both research and development implications. A rough definition of plant biotechnology is offered:

> *An applied field of science whereby scientific principles are used to discover new methodology and instrumentation to produce new forms of and uses for biological entities.*

Obviously, the discovery phase has a research orientation and the production phase is a technology development and transfer process. Plant scientists may function in both phases, and are therefore *plant biotechnology research scientists.* The second phase has created a new professional area and opportunities for

In assessing the progress of plant biotechnology it must be clear that there are both research and development implications.

plant biotechnologists. The land grant universities have responsibility for research, training of plant biotechnologists, and, to varying degrees, product development. The consumer community perceives that the research universities will be involved in both phases and will deliver products for their use. This is unrealistic in most cases because the prevailing philosophy among universities has been that product development can best be done by the private sector. Countering that, however, is the fact that State Agricultural Experiment Stations (SAES) and the USDA have developed, evaluated, released, and distributed crop cultivars and germplasm for many decades as a public service, and, in fact, a responsibility. The major U.S. crops include public-developed germplasm in their breeding history or are direct products of public-supported research and development programs.

There is some uncertainty administratively and by scientists about how far the universities' research and development programs should take their discoveries. From the administrative point of view there are philosophical issues about the university role and private sector role. In the past it has been rather clear that the private sector should do developmental research and market products to the extent possible. In the case of cultivar development, the private sector incentives are not sufficient for all but a few crops, so the public sector assumes responsibility. More recently, however, administrators have found it increasingly difficult to amass the financial and physical resources to accommodate research programs, especially for high-cost biotechnology research. Therefore, it is attractive to consider cost-recovery mechanisms through development of products for direct sale or licensing. This obviously impacts on the scientists who must balance personal goals and university advancement criteria, with the financing of a research program. The days of SAES providing sufficient research facilities and operational costs for research and development are past. The scientist must supplement the provided funds with funds from other sources. A moderately active laboratory for a plant biotechnology program requires at least $1,000 per day of extramural funds to meet operational costs and even more when indirect institutional costs are transferred to the investigator.

...administrators have found it increasingly difficult to amass the financial and physical resources to accommodate research programs, especially for high-cost biotechnology research...

This is a heavy burden to place on a scientist who must teach classes, serve on committees, and train graduate and postdoctoral students. Research funds are presently mostly available through competitive peer-review awards. These are most successfully obtained as individual-scientist awards and only a few opportunities exist for funding multidisciplinary research teams for a comprehensive research and development program. Furthermore, almost no competitive grant funds are available for the product-development phase. This has favored short-term research on specific topics which yield new knowledge, appropriately published in peer-review journals, and has not favored long-term projects which will produce new genetic stocks and products ready for consumer use. Thus, the directions of university research have been dictated largely by the individual investigator's ability to be successful in competition for extramural funds.

...almost no competitive grant funds are available for the product-development phase.

There have been several creative joint ventures of private sector and university research, based on shared knowledge and biological products, usually with first right for licensing being obtained by the private sector entity. University scientists have entered these agreements willingly because the personal risks were minimal with respect to product development and licensing and also because there were minimal restrictions on publishing and patenting. From the private sector side I suspect that the ultimate payoffs in licensable technology have been minimal, but the information exchange and training elements were very useful. In some cases known to me the private sector funding was short term, three to five years, so that realistically little plant biotechnology research could reach the product-development stage.

From the above comments it is clear that there are institutional and scientist-level considerations that impact on progress in plant biotechnology. There are real impediments borne by the university scientist. The typical institutional response on the research financing issue has been to write a statement into position descriptions that "the successful candidate will be expected to generate research funds through competitive grants and other means," notwithstanding other expectations to "serve agriculture and the general public by discovering new knowledge and developing new materials."

GOALS AND PRODUCTS OF PLANT BIOTECHNOLOGY

This is a heavy burden to place on a scientist who must teach classes, serve on committees, and train graduate and postdoctoral students.

The definition of plant biotechnology offered earlier really is not unique to the present era of plant science research and development and it could have appeared any time in this century. However, scientific developments of the 1960s and 1970s have provided a whole new set of tools for applied biological sciences. Research goals in plant biotechnology are likewise not unique to the present time and include, among others, the following:

—New fundamental knowledge about plant processes, including gene structure, function, and expression.

—New methodology for application in product development.

—New genetic stocks, including plants, gametes, and cloned DNA, for research and plant breeding.

—New cultivars for direct use in agriculture.

What is unique to the present era and is so exciting for the advancement of agriculture, food quality and safety, energy, and health care products is the

I believe that neither the faculty-reward incentives nor the funding sources have encouraged long-term research and development efforts.

practically unlimited number of product-design topics that can now be addressed by manipulations of biological materials. Examples are included in Table 1. Most often gene transfer is highlighted as the cornerstone of plant biotechnology and certainly the possibility of unrestricted interspecific gene transfer is a major development. However, other developments leading to precise, rapid, and low-cost diagnosis of the presence of disease-causing entities or toxic compounds in plants have far-reaching implications for U.S. agriculture and monitoring of consumer product safety. Another development that requires little or no manipulation of genetic materials is rapid clonal reproduction of disease-free plants. Cell, tissue, and organ culture methods have already been usefully adopted for commercialization of cultivars and for safe international transfer of genetic resources for research purposes.

Few products of the type listed in Table 1 have emerged from university research and development, but surely these institutions have advanced new scientific concepts and methodologies. While we are still perfecting technology to accommodate product development, I believe that neither the faculty-reward incentives nor the funding sources have encouraged long-term research and development efforts. There is also a trend toward 9-month appointments in the SAES which may also discourage long-term research and development. There are signals through press releases that a great increase in product-oriented research is being undertaken at some universities. This is encouraging because it suggests that stable funding for biotechnology may be emerging as it was in the past for "agri-technology"—traditional agricultural research in the SAES.

GENETIC RESOURCES

By definition, biotechnology requires biological materials and thus is absolutely dependent upon biological resources. Biotechnology applied to plants uses biological resources from microbes, animal, and, of course, plants routinely. Specific genes or nucleotide sequences are the basic genetic resources required not only for gene transfer but also to adopt enzyme systems for manipulating DNA. A central dogma is that any gene from any species may be transferred to any other species by molecular manipulations and parasexual means. In practice this means that the entire global gene pool is accessible for introducing genes into specific plants. Genes from plants have been transferred and expressed in nonplant species as well.

Clearly, genetic resources are fundamental to plant biotechnology research. Many plants have been collected from their native habitats and amassed in seed banks or gardens. These efforts have been under way for more than one century in the development of the U.S. genetic resources which were obtained from

TABLE 1 SOME PLANT BIOTECHNOLOGY PRODUCTS

1. Cloned genes for insertion into cultivars
   - a. genes integrated into plant genome
   - b. gene therapy
2. Diagnostic methods
   - a. to detect pathogen infection
   - b. to detect presence of genes in plants
     - linked markers
     - DNA–RNA complementation
   - c. to assess genetic identity
3. Rapid clonal propagation of plants
4. Gene banks of cloned DNA or nucleotide sequence data
5. New plant products for commercial use
   - a. genes transferred from another species
   - bnthesized genes
6. Plant products produced by microorganisms

By definition, biotechnology requires biological materials and thus is absolutely dependent upon biological resources.

practically all countries. The central dogma here is that "genes are free," and since only small samples of such renewable natural resources have been collected and taken away from their natural habitat, there was little concern that plant collectors were adversely affecting the economy or ecology of any country. There were a few exceptions, however, where economical considerations dominated, for example, for the plants contributing rubber, coffee, and tea.

More recently two issues have emerged. The first is that ancillary development factors—new agricultural systems, crop substitution, water development, excessive animal grazing and extractive harvests, urban expansion, and industrial wastes toxic to terrestrial and aquatic species—have threatened the integrity of the native gene pools. The second is that the "genes are free" dogma has been challenged by some genetic resource-rich countries and social scientists. Genetic resource conservation by *in situ* and *ex situ* means must be given much higher priority in view of the difficulty to manage the development factors mentioned above and because only planned biological conservation can provide assurance that genetic resources will be available and accessible for the indefinite future.

Plant biotechnology depends upon genetic resources but also restricts the free exchange of materials. Ownership of genes is an especially important concept. A genetic resource obtained "for free" may yield, upon extensive manipulation and characterization, a valuable gene or gene complex or result in a new process for manipulating genes. Then the "genes are free" concept breaks down,

...it is unlikely that genetic erosion of agricultural species will be enhanced by introduction of methods from biotechnology.

especially if there is commercial potential in the derived genetic materials. The issues of ownership of genes and ability to protect their use through patenting and other processes are contentious. Further study is essential.

Plant biotechnology practitioners, public and private, at the policy and research levels must give high priority to genetic resource conservation, and to the "genes are free" or "genes for a fee" concepts. With respect to genetic resource conservation, molecular methods have been rapidly adopted for analysis of genetic diversity which has aided in developing *ex situ* and *in situ* conservation strategies. The potential impact of plant biotechnology in reducing or enhancing the genetic resource base remains unclear; however, it is unlikely that genetic erosion of agricultural species will be enhanced by introduction of methods from biotechnology.

PLANT BIOTECHNOLOGY AND PLANT BREEDING

The greatest hope and hype in plant biotechnology remains the potential for unrestricted gene transfer. This possibility is very attractive to plant breeders because it permits directed gene transfer and because the genetic resource base for a targeted crop species is vastly expanded.

Crop development through plant breeding is conceptually simple: locate a gene resource; introduce it to a population of plants; select plants having desirable combinations of traits; evaluate these plants for field performance, consumer acceptability, and safety; release the derived cultivar for general use; and, finally, arrange a distribution and marketing program. Plant biotechnology does not alter this sequential process at all, but it does present the plant breeder with new opportunities and challenges at each step in the cultivar-development process. Table 2 lists eight topics in plant breeding which are influenced by plant biotechnology. Some brief comments are offered because it is the analysis of these topics that gives a clearer picture of constraints and bottlenecks in transferring new technology to crop development.

TABLE 2 COMPONENTS OF PLANT BIOTECHNOLOGY IN CROP CULTIVAR DEVELOPMENT

1. Genetic resources
2. Trait identification
3. Gene isolation
4. Parasexual gene transfer from donor to recipient cultivar
5. Gene expression
6. Breeding—creating adapted gene complexes
7. Evaluation for agronomic performance and safety
8. Distribution and marketing of seed stocks or plant propagules

1– *Genetic resources*—Availability of germplasm from which genes may be extracted is quite good for major crops, but less so for wild relatives. Perhaps the largest bottleneck is that the vast holdings of crop germplasm in the U.S. National Plant Germplasm System are largely undocumented or not evaluated for potentially useful traits. Thus for biotechnologists a first step is to obtain a germplasm collection that may be evaluated for traits of interest. The fact that most genetic resource collections are inadequately characterized is not a surprising deficiency because many traits being considered for transfer were not even recognized as valuable in the past.

2– *Trait identification*—Which traits are to be selected for transfer is a somewhat contentious issue. Some traits may have potential for commercial exploitation but are not necessarily desirable from other points of view; herbicide-resistant plants is an example thoroughly discussed elsewhere in this volume. Biotechnology is geared to transfer of single genes, and there are many important single-gene traits as candidates, pest resistance and crop quality factors are examples. However, the adaptation of a crop to a production environment requires combinations of many traits, most of which are multigenically controlled. Thus, transferring genes one at a time for multigenic traits is not feasible.

The alternative approach is to identify chromosome regions which have blocks of genes that contribute to the desired trait expression. Thus, it is not necessary to identify the specific gene for a trait; linked genes or nucleotide sequences can be used as markers of the desired gene complexes. DNA restriction fragment length polymorphisms (RFLPs) have provided such a tool in trait identification that is probably equally as important as single-gene identification in plant breeding. Medium-to-high resolution genetic linkage maps are needed for this purpose. These maps are conceptually easy to develop, but expensive in time and laboratory supplies and require highly developed genetic stocks for some crop species. Facilitating linkage map development is a top priority for plant breeding purposes.

3– *Gene isolation*—Methods to probe for DNA sequences for particular genes are available, but more efficient means are necessary for routine plant breeding use. Particularly useful are congenic isolines which plant breeders have developed for many traits. However, these are underexploited for gene identification purposes. Breeders need access to cloned genes, and the process of development of such genetic stocks could be enhanced by multidisciplinary teams (molecular biology, genetics, and plant breeding).

4– *Parasexual gene transfer*—To transfer single genes, methods for introduction and integration (transformation) of cloned DNA into plant genomes is essential. The main contribution of biotechnology to plant breeding is to eliminate "linkage drag," that is, the transfer of undesirable genes along with desirable

Evaluation of a genetic resource collection for a trait of interest yields information at the phenotypic, but not genotypic, level.

ones. Traditional plant breeding relies on random genetic recombination and selection to eliminate such association. True single-gene transfer eliminates one of the most costly and time-consuming efforts in plant breeding. Up to now gene transfer methodologies have not been available for routine use in plant breeding, and, in fact, only primitively developed for grass species. This is clearly a major constraint in single-gene plant breeding.

5– *Gene expression*—Evaluation of a genetic resource collection for a trait of interest yields information at the phenotypic, but not genotypic, level. Therefore it must be shown by appropriate tests that there is a genetic component to the observed morphotype. Most important, however, is to determine if the trait is monogenically or multigenically controlled since this will dictate a gene-cloning strategy and have bearing on the practicality of attempting to transfer the trait to a crop cultivar.

Expression of a transferred single gene must also be assessed in the progeny because it is widely known from experiences in plant breeding and genetics that the genetic background (i.e., genes on other chromosomes) may affect the degree of expression of a single gene. Experience with some species have shown that the transformed plants lack vigor and are modified somehow by the gene-transfer process. This phenomenon needs considerable attention from the developmental genetics approach. Gene expression can be a considerable constraint for plant breeding, but the extensiveness of this problem is not known.

6– *Breeding-creating adapted gene complexes*—Crop cultivars are the product of many generations of selection, and farmers' landrace varieties provide ample evidence for this. Each plant breeding program must retain or reconstruct the desirable combinations of genes in new cultivars. Thus single-gene transfer routinely applied can be a great benefit to breeding. However, most breeders would consider these genes to give an expansion of their gene pool and would incorporate them into breeding populations for recombination with other genes controlling many other traits.

The new technology offers the possibility of selecting for groups of genes through the RFLP technology mentioned above. Thus biotechnology can provide both genes and gene-tagging methodologies which will aid the plant breeding process. RFLPs are not sufficiently developed for use in breeding, and this activity needs high-priority attention because the resulting DNA probes and linkage maps will be of general use for any plant breeding program.

7– *Evaluation for agronomic performance and safety*—This is another contentious area for plant breeders because there is concern that new traits introduced into plants may have undesirable effects. Plant breeders have contended with this consideration continuously and have developed the necessary evaluation techniques to assure that a new cultivar was not toxic or created no problems as a weed, thus informally plant breeders have found the GRAS – General Recognized As Safe – concept to serve public needs adequately. The guiding principle outlined in a National Academy of Sciences report (Kelman report) is that it is the trait and its effects which must be evaluated. The process by which a trait is introduced to a plant does not create a new potential for undesirable effects. This view was substantiated by a detailed National Research Council report in 1989. A decision tree to guide testing and containment procedures was developed, and these procedures, for the most part, are being adopted in federal agency guidelines. Some breeders, and I am one of them, are pleased to have oversight guidelines for testing and evaluations, but do not believe regulatory language is needed at the federal level for what is or will become a routine procedure in agriculture. Oversight on food safety of new products is well established and requires no new regulatory practices.

...plant breeders have found the GRAS – General Recognized As Safe – concept to serve public needs adequately.

This is not to say that biotechnology does not create any new potential problems. It does require that the gene and its effects must be carefully assessed. For example, gene transfer by hybridization to crop relatives has been considered in the past, but is the consequence the same for a seed protein quality gene and a herbicide resistance gene? Questions of this type are routinely considered in the plant breeding process and will continue to be addressed by appropriate experimentation. The discussion on this point should continue and crop by crop analyses are very appropriate as is currently in progress by the United States Department of Agriculture/Animal Plant Health Inspection Service.

8– *Distribution and marketing*—New crop cultivars may contain patented genes, and the cultivar itself may be protected by a plant patent, a plant variety protection certificate, or various forms of trade secrets. This is not a new development, but there is increased uncertainty about how developers may protect their investments and realize profit for their efforts. Some crops may not receive sufficient attention if the questions about protection of intellectual property are not addressed. This may be most important for some self-pollinating species such as wheat, soybeans, and cotton. Public-sector crop developers need also to take full consideration of these issues so that they can be assured that the products of their efforts will be readily available to the general public. New arrangements for selective licensing and patenting to prevent exclusionary uses of new cultivars are both instruments that may be applied to the advantage of the consumer depending on the crop, its area of adaptation, and extent of usage.

Presently plant breeders are wary of adopting some biotechnology practices,

...there is increased uncertainty about how developers may protect their investments and realize profit for their efforts.

such as gene transfer by DNA on projectiles, because they are unsure how a cultivar could be released if its development included the use of a patented procedure.

PLANT POPULATION BIOLOGY

The impact of plant biotechnology on the discipline of plant population biology and the study of evolutionary processes on plants is becoming well-known. Molecular methods for assessing allelic diversity are being routinely adopted for study of mating systems, population differentiation, and effects of natural selection on populations of many species.

Another aspect is that gene transfer to crop plants may profoundly alter the genetic structure of sexually compatible species. This is a concern only now being addressed theoretically and experimentally. Decisions may be made about whether a gene from a crop will be significant if transferred and incorporated into wild, sometimes weedy, populations. These discussions will be made based on biological, economic, and social criteria. It is clearly an area of uncertainty for plant breeders, but one which can be solved on a case-by-case basis. Especially interesting is the potential for genes to be introduced by crops into the centers of origin and diversity for a crop species. This requires further study.

WHY HAS PLANT BIOTECHNOLOGY NOT HAD A GREATER IMPACT?

The elementary tools for plant biotechnology were developed in the '50s and '60s (tissue culture) and '70s and '80s (gene cloning and transformation). For the most part, gene transformation is still in the research phase. Thus, methodologies are still evolving. Next, the application of these methodologies to produce a cultivar, followed by its testing, multiplication, and release require about the same amount of time as do conventionally derived cultivars. This is about six years. So for a cultivar to be released in 1991, a potentially new cultivar would have emerged from the laboratory in about 1985. Prior to that all of the tools of tissue culture, gene cloning, and transformation had to be in place and used. In the early 1980s only a few systems were workable. Thus, I conclude that biotechnology is not lagging in cultivar development, because the tools plant breeders use are still being assembled. For other uses, diagnostics, pest control, and clonal plant propagation, for example, there are successful products being marketed, and this trend should accelerate in the near future. There are many issues to discuss and research, but the future of plant biotechnology, when integrated into agricultural and health sciences looks very promising.

...biotechnology is not lagging in cultivar development...

USE OF MICROORGANISMS FOR CROP AGRICULTURE

Since the turn of the century there have been many research programs, worldwide, attempting to develop cultures of microorganisms useful for crop agriculture. However, there are relatively few examples of such inoculants being used on the farm. In comparison, pure cultures of many types of microorganisms have been very important to the pharmaceutical and food industries, which continually genetically alter strains to improve them. Of course, these latter industries grow microorganisms under controlled conditions such as a fermenter for antibiotic production or temperature-controlled milk for yogurt or cheese. In contrast, microbial inoculants to be used by a farmer have to exert their positive effect under tremendously variable field conditions, such as weather, soil type, plant variety and field history.

WINSTON J. BRILL

Winston J. Brill & Assoc.
4134 Cherokee Drive
Madison, WI 53711

Founder of Agracetus, Brill is now principal at Brill & Associates, an organization designed to assist corporations improve research productivity and creativity.

In the first decade of the 1900s, farmers in Europe and U.S. became very interested in a recently discovered bacterium, called *Rhizobium*, that dramatically increased yields of legumes such as soybean, bean, pea and alfalfa. These bacteria form nodules on legume roots and convert nitrogen gas from the air to ammonia, which is used by the plant. Thus, such inoculated plants no longer require addition of nitrogenous fertilizers, such as nitrate, to obtain high yields. Through this practice, nitrogen was added to the soil in a manner that prevented it from polluting bodies of water, through run-off. Many commercial *Rhizobium* inoculants have been marketed since then. Thousands of field tests have been performed worldwide in order to determine which *Rhizobium* strains are the best for a particular plant variety in a specific growing region. These tests were performed in university, government and commercial settings.

Success with *Rhizobium* stimulated laboratories to search for other types of microorganisms with the potential to aid agriculture. In the 1920s

...it seems that certain microorganisms actually do stimulate crop yield and/or pest antagonism.

and 1930s, the literature had hundreds of examples, from a wide variety of microbial species, that seemed to stimulate growth of a plant or protect the plant from pests such as insects or fungi. This type of work has continued to the present. However, only a few strains are currently marketed. *Bacillus thuringiensis* is the most commonly used example.

Many results describing potentially useful microorganisms were either inconsistent or they were not reproduced by other researchers. In the U.S.S.R. during the 1950s about 25 million acres were inoculated with bacteria such as *Azotobacter chroococcum* and *Bacillus megaterium* for a wide range of crops, including potato and wheat. While the popularity of these inoculants has substantially decreased, papers continue to be published on limited demonstrations of efficacy. Many of these reports come from credible and sophisticated laboratories. So it seems that certain microorganisms actually do stimulate crop yield and/or pest antagonism. However, parameters that influence the effectiveness of the microorganism are not understood, or are not controllable, and yield increases usually are quite sporadic.

A major hurdle to overcome for developing useful inoculants is that the microorganism usually does not persist in high concentrations for a sufficient length of time to affect the plant in a positive manner. That is why much of the research focuses on the germination stage of the plant. To influence germinating seeds, (e.g., through microbe-produced plant-growth hormones, or microbes which inhibit fungi that cause seedling damping-off diseases), it is relatively easy to apply high numbers of the inoculant to the seed. High populations of the microorganism can be added directly to the seed coat at the time of planting. However, as the plant develops, the number of inoculant microorganisms in contact with the plant dramatically decreases, and the inoculant rapidly loses its effectiveness. To overcome this problem, researchers are looking for strains that bind to the plant (e.g., to roots) and may, therefore, multiply during plant development. So far, this has not been successful; thus, beneficial effects are transitory—usually occurring shortly after plant or soil inoculation.

With the excitement about biotechnology in the late 1970s and early 1980s, interest in microbial inoculants was stimulated. There seemed to be tremendous potential to develop new types of agriculturally useful products through these modern technologies. It is relatively easy to isolate genes of interest, such as those that code for pest antagonists, those that produce plant growth hormones or those that degrade unwanted organic chemicals. It is quite easy to add genes to most microorganisms. Also, it seemed that it should be relatively easy to continually improve products through genetic alterations—as was the experience in the food and pharmaceutical industries. Microorganisms have the potential to be more environmentally compatible than many chemicals used in agriculture. Inoculant practices may play an important role for sustainable agriculture. These incentives induced some large chemical, agricultural and pharmaceutical companies to initiate inoculant research programs. A number of small start-up companies also focused on this area. In the past 15 years, there have been many examples of significant and reproducible plant growth stimulation, yield increase or pest inhibition in greenhouse and growth-chamber studies. However, most of these companies have now completely eliminated these programs. What happened? Promising results were not observed from initial field trials.

Microorganisms have the potential to be more environmentally compatible than many chemicals used in agriculture.

Many of these projects were terminated prematurely. Most of the scientists working on these programs did not keep the complexity of the field in mind during all stages of the project. Thus, excitement from greenhouse or growth–chamber results was frequently dampened when the organism was field tested. Agronomists with extensive field experience know that greenhouse and growth-chamber data most commonly do not relate to what occurs in the field, with all of its variability and complexity.

Steps in a research project...all should be analyzed in the field.

Field tests are essential from the earliest stages of a program to develop microbial inoculants. If this is ignored, then there is a great chance that laboratory work will be a waste of effort. Steps in a research project, such as optimizing the growth medium, genetically altering the strains and formulating the microorganism, all should be analyzed in the field. The simple activity of isolating strains can induce unwanted mutations that could keep the microorganism from being effective. The organism may behave like the parent strain in the laboratory, but in the soil, for example, it may be hypersensitive to dry conditions that may be faced in the field. Mere scale-up of a growth medium from small flasks to larger vessels may render the microorganism physiologically inactive for its beneficial property. Experience in the pharmaceutical industry has demonstrated that problem on many occasions. Thus, extensive field tests are essential at each

How can such a research project be pursued if efforts to secure required (or recommended) data and documentation to satisfy regulatory agencies, for even the smallest field test, costs several hundred dollars?

step in developing a microbial inoculant. In fact, data usually become meaningful only when field tests are performed at several sites and sufficient replicates permit useful statistical analyses.

There may also be opportunities to increase crop yield by specifically breeding plants for enhanced effectiveness of the inoculant. This has been demonstrated with the legume-*Rhizobium* partnership as well as other experimental inoculants. Modern breeding may have removed genes important for maximizing the plant-microorganism association. Of course, extensive field trials are the only way to optimize plants for inoculants.

Unfortunately, the current regulatory situation, for field tests with genetically altered organisms, has resulted in a disincentive for university, government and industrial researchers to pursue microbial inoculants. With these new regulations and guidelines, organisms modified by traditional genetic methods are to be included with organisms modified by recombinant DNA methods, since it seems to be agreed by both researchers and regulators that a recombinant organism *per se* should be no more dangerous than the same organism modified by older, less precise, methods. An outdoor test of even one square foot must be scrutinized in enormous detail before permission is granted. How can such a research project be pursued if efforts to secure required (or recommended) data and documentation to satisfy regulatory agencies, for even the smallest field test, costs several hundred thousand dollars? Note that almost all microbial field tests, so far, have been sponsored by corporations. These companies now have become more wary of greenhouse or growth-chamber results. So, it will be even more difficult for university scientists to find a sponsor for an early field test to try out an idea. Meager federal research grants cannot support work to satisfy regulators.

Many investigators now realize the importance of field tests at early stages of an inoculant program, but very few can handle the regulatory burden of a research program that allows, for example, an interesting microorganism developed through modern or traditional genetic techniques, to be field tested in different types of fields at the earliest stages of the program. Thus, a research area with high potential to help agriculture and the environment has been considerably slowed.

Certainly, research that has a reasonable chance to damage health or the environment should be tightly regulated. It seems that current regulations and guidelines assume that genetically altered microorganisms have reasonable potential to be harmful. This does not make sense—based on a century's worth of extensive experience with field tests of wild-type and genetically altered microorganisms. The most sophisticated technique, genetic engineering, adds a characterized gene to the microorganism. Other techniques, such as mutation or plasmid transfer are less predictable than genetic engineering, as far as the properties of the microorganism are concerned. It is well known that mere isolation of a microorganism from the soil will add uncharacterized mutations; thus, each wild-type microorganism (possibly thousands of strains) field tested since the turn of the century had "uncharacterized genetic mutations." We have yet to hear of a single health or environmental problem resulting from this type of research. Previously, many microorganisms with laboratory-directed mutations or with genes added by natural plasmid transfer have been field tested without any reported untoward effects—or regulatory concerns.

In comparison with agricultural microbial research, agricultural chemical research routinely involves small field tests without regulatory scrutiny.

Compare the difference between adding a genetically altered microorganism to a field versus adding an experimental chemical to a field. When a chemical pesticide or fertilizer is added to soils, it is known that certain mutations and gene transfers by indigenous microorganisms are greatly enriched. Most of these microorganisms and/or their genetic alterations are "uncharacterized." Ecological experiments continually demonstrate natural gene transfers between different genera in soils and bodies of water. So, microbes with genetic changes in chemically treated fields most probably transfer their altered genes to different genera and species. We know of no health or environmental problem that has occurred from these uncharacterized organisms with uncharacterized genetic changes.

Evolutionary principles govern microbial populations and persistence. While microbes added to a field rapidly decrease in numbers, some chemicals persist for a long time. In comparison with agricultural microbial research, agricultural chemical research routinely involves small field tests without regulatory scrutiny. As a chemical exhibits applied promise, after analysis of many small field tests, then regulatory approval is necessary to advance to large field tests and possible commercialization. So, the chemist's initial field experiments are unhampered by regulators, while the microbiologist's initial field experiments require extensive regulatory

Current regulations and guidelines strongly inhibit advancement, while not really protecting our health and environment.

scrutiny. Both types of tests cause "new" microorganisms to develop. It is assured that these microorganisms will be transported out of the field by such factors as the wind and insect movement.

Some commercial inoculants may be ineffective with certain future crop varieties. Also, there may be circumstances in which an inoculant product actually decreases yield. Such situations do not cause environmental or health problems. The commercial value of the product merely decreases. These types of problems have been, and continue to be, found with some commercial agricultural chemicals.

The chance of experiments, aimed to help agriculture, unintentionally converting a harmless microorganism to one that damages health or the environment seems to be exceedingly small. The chance that current regulations and guidelines will detect this very rare event also seems to be exceedingly small.

Microbial inoculants have potential to increase crop yield without damaging the environment. If regulations would be based on scientific knowledge and would consider our extensive experience with genetically altered microorganisms, we may be able to advance the microbial inoculant research area and make concerted efforts to solve some important agricultural and environmental problems. However, current regulations and guidelines strongly inhibit advancement, while not really protecting our health and environment. Hopefully, agencies will eventually design regulations appropriate for research and commercialization. A balance must be made between protecting the public from problems, and helping the public benefit from potentially desirable agricultural practices. Unfortunately, we have seen little progress towards that balance.

BIOTECHNOLOGY IN ANIMAL AGRICULTURE

GEORGE E. SEIDEL JR.

Animal Reproduction and Biotechnology Laboratory
Colorado State University
Fort Collins, CO 80523

For over twenty years, Seidel has worked to perfect techniques for manipulating embryos. Current projects include procedures to make transgenic animals, in vitro oocyte maturation and fertilization, chemically defined media for culturing embryos and genetic control of maternal recognition of pregnancy.

I have been asked to assess the interface of biotechnology and animal agriculture, paying particular attention to recent successes and failures, and to identify potential opportunities as well as bottlenecks to further advances. Some aspects of this assignment are relatively straightforward, while others require so many qualifiers as to be of limited value. The first problem is to define biotechnology. In a broad sense, every aspect of agriculture is biotechnology, whereas a narrow definition limited to recombinant DNA techniques would show only barely perceptible effects on production agriculture at this time. A broad definition of biotechnology will be used and items of potential interest will be emphasized. Defining animal agriculture is also somewhat arbitrary. Currently, the most important agricultural species in the United States are cattle, swine, sheep, chickens and turkeys. Horses might be considered more recreational than agricultural. A considerable percentage of beef cattle operations also have a huge recreational component, as do those concerning other species. Many of the opportunities to apply biotechnology are on farms where the majority of family income is from non-farm sources. There also are biotechnology applications in less common species such as buffalo, deer, goats, ducks, geese and fish.

BRIEF SUMMARY OF THE STATUS OF ANIMAL AGRICULTURE

Currently about two percent of the labor force in the United States is employed in production agriculture, and another 20 percentof the labor force is involved in servicing production agriculture (e.g., fertilizer, farm equipment) or transporting, processing, and marketing agricultural products. Roughly half of these people are in the animal agriculture sector. This

In a broad sense, every aspect of agriculture is biotechnology...

huge enterprise is reasonably successful by most measures, e.g. providing abundant, healthful products at the lowest cost in the world as measured by percentage of median disposable family income used for food (currently 12 percent). This is all the more remarkable because of the huge percentage of food expenditures that go for packaging, storage, processing, safety and quality control, advertising, and service in restaurants and related establishments.

This situation is less rosy when examined in depth; e.g., those involved in production agriculture have not fared well over the years (a robust measure of this is the continuing exodus of people from production agriculture, particularly involving animals). The huge direct and indirect government subsidies to address this problem have met with only marginal success and add indirectly to the cost of food, although these costs are low compared to most other industrialized countries. It also must be made plain that improving efficiency of animal agriculture with biotechnology, while greatly benefitting consumers, will probably harm more farmers than it will help because in our economic system, those who do not adopt more efficient techniques cannot compete with those who do. On the other hand, unless American farmers become more efficient, their jobs will be replaced by farmers in other countries who export their produce to the United States.

...unless American farmers become more efficient, their jobs will be replaced by farmers in other countries who export their produce to the United States.

BIOTECHNOLOGIES ARE TOOLS

Agricultural practices are continually refined and perfected. For example, 10,000,000 dairy cows currently produce considerably more milk than 25,000,000 did in 1945; in 1990 beef production from calves of 33,000,000 beef cows exceeded that of 45,000,000 cows in 1975. Similar successes have been recorded for poultry and swine. There are huge opportunities to improve this performance further, particularly with cattle, since the best herds produce twice the output per animal as the average herd. Continued improvements arise from dozens of sources, each of which only contributes modestly. An excellent example is selective breeding. Although the improvement with each generation is modest, it accumulates like compound interest. In most cases, new biotechnologies should be thought of as tools to aid this process. They face stiff competition from current tools, which already are very effective. Any compromise of currently effective tools must be compensated by extraordinary benefits, or the new technology will not even be considered. A good example is introducing a transgene for disease resistance into a breeding population of dairy cows. Unless the transgene is already in a high milk-producing line, the economic costs of lowered milk production in the initial generations would cancel the benefits of controlling any known disease.

RECENT BIOTECHNOLOGICAL BREAKTHROUGHS

I have constructed three lists of biotechnologies used in production animal agriculture (Tables 1–3). The first list, technologies commercialized before 1980 (Table 1), might be considered inappropriate for the objectives of this paper. Nevertheless, I present them because they are of overriding

TABLE 1 EXAMPLES OF BIOTECHNOLOGIES IN WIDESPREAD USE COMMERCIALIZED BEFORE 1980

Selective breeding
Nutrient requirements of animals
Feed analysis
Vaccination
Veterinary diagnostics and therapeutics
Artificial insemination (mostly cattle and poultry)
Crossbreeding
Regulation of reproductive cycles
Embryo transfer
Ultrasound to measure carcass fat

importance in current animal agriculture, they illustrate the kinds of technologies that work, and some of them just came into widespread use around 1980, e.g., regulation of reproductive cycles and embryo transfer technology. It is important to recognize that most technologies are applied directly only to a minority of animals in the population. Some of these niches only apply to specialty markets, e.g., embryo transfer to introduce new germ plasm into specific pathogen-free swine herds, while others filter down to the whole population, e.g., artificial insemination is only done with five percent of beef cattle, but over half of the bulls that breed beef cattle via natural service are conceived by artificial insemination or had parents conceived by artificial insemination. Some biotechnologies apply to every individual in the population. For example, all turkeys are conceived by artificial insemination because males have so much breast meat that it interferes physically with natural mating.

The second list, biotechnologies commercialized in the 1980s, (Table 2) similarly has technologies widely applied, e.g., growth promotants for growing beef cattle, and those used in narrow niches, e.g., storage of bovine embryos in liquid nitrogen. This latter technology resulted in an entirely new commodity, frozen embryos. Many frozen embryos are imported and exported; a very important fringe benefit of this approach is the greatly decreased danger of spreading diseases compared to importing live animals or semen.

TABLE 2 EXAMPLES OF BIOTECHNOLOGIES IN WIDESPREAD USE COMMERCIALIZED AFTER 1980

Growth promotants (other than DES) and ionophores
Use of bypass (of rumen) protein and fat
Monoclonal antibodies for diagnostics
Storage of embryos in liquid nitrogen
Ivermectin to treat parasites
Ultrasound for veterinary diagnostic purposes

Ivermectin is truly a miracle drug in terms of safety and efficacy. Although this drug was developed for animals, and there is no use for it in people in the United States, it will be one of the most important drugs ever developed for human use in countries where people are regularly infected with certain deadly or debilitating parasites, e.g., the organism causing river blindness in Africa.

The third list, (Table 3) technologies just being commercialized at this time, will be discussed item by item. The list is headed by bovine somatotropin (a growth hormone), a truly marvelous product of biotechnology produced by bacteria via recombinant DNA technology and administered to dairy cows by injection or implant about every two weeks. This treatment increases milk production about 20 percent compared to controls. For sound reasons, it is not administered during the first quarter of lactation, so the increase in milk production is closer to 10 percent over the lifetime of the cow. The main action of this protein hormone in lactating cows is simple: it is a so-called partitioning agent, causing the nutrients that a cow eats to go to milk production rather than carcass fat. Without exogenous somatotropin milk production declines markedly in many cows after the first third of lactation, and many of the nutrients they eat end up as body fat. Somatotropin has other effects too, e.g., it stimulates appetite so cows eat more, thus further increasing milk production.

This product (somatotropin) is safe, efficacious, cost-effective, and does not alter milk composition, yet it is one of the most controversial products ever developed.

This product is safe, efficacious, cost-effective, and does not alter milk composition, yet it is one of the most controversial products ever developed. One problem is current milk surpluses; another is that it is a recombinant DNA product. However, it is unlikely that even a few percent of recombinant DNA products will be able to come close to the safety and efficacy parameters of this product; it should be a great cause of concern that this product is perceived so negatively by so many.

At this writing there is a glut of crude oil on the world market, but think of the benefits of 10 percent increase in fuel efficiency of internal combustion engines. There would be costs too, less employment in oil fields just as fewer dairy cows and dairy farmers will be needed after bovine somatotropin comes into use. However, precisely these kinds of efficiency are required for the American dairy farmer to compete; also, a lower number of cows equates to less pollution, less grain consumption, etc.

TABLE 3 EXAMPLES OF BIOTECHNOLOGIES BEGINNING TO BE USED COMMERICALLY IN THE EARLY 1990S

Bovine somatotropins to improve milk production
Vaccines inducing antibodies distinguishable from those due to natural infections
Cloning embryos by nuclear transplantation
Sexing embryos
In vitro maturation of oocytes from slaughterhouse ovaries
In vitro fertilization and culture of embryos
Oral melatonin to control seasonal breeding in sheep and goats
Induced twinning in cattle

The second item in Table 3 is a solution to an old problem, distinguishing healthy animals with antibodies due to vaccination from animals making antibodies due to current infection with a contagious agent. A well-known example is bovine brucellosis, which causes great economic loss due to abortion. This organism also is a threat to people; it causes undulant fever, which usually cannot be treated effectively, and is very de-

...making exact genetic copies of outstanding animals by nuclear transplantation has been a dream of animal breeders for years.

bilitating for years. The program to eradicate brucellosis in cattle has been extremely expensive and hindered by the above problem. However a new vaccine developed with recombinant DNA techniques is about to be released that elicits antibodies distinguishable from those induced by the brucella organism itself. A similar vaccine has been developed for pseudorabies in swine.

The third item in Table 3, making exact genetic copies of outstanding animals by nuclear transplantation has been a dream of animal breeders for years. This now can be accomplished by combining serial cloning of embryos (e.g., transplanting nuclei from a 16-cell embryo into 16 unfertilized oocytes, and when they reach the 16-cell stage, repeating the process) and cryo-preservation of embryos so that a genetic copy in embryo form is available after other cloned embryos become adults. If the adults are outstanding, copies can be made; if not, the embryos are discarded and other clones propagated. This technology has other advantages: automatic sex selection, faithful reproduction of heterozygous transgenic animals, and reproduction of outstanding crossbred animals. Note that without this technology these latter two types of animals do not breed true, much like the situation with hybrid corn.

Item four is sexing embryos. Although still a bit clumsy, commercial sexing of bovine embryos with a Y-chromosome-specific DNA probe is now offered commercially. This technology eventually will be replaced by sexed semen. The fifth item in Table 3, *in vitro* maturation of oocytes from slaughterhouse ovaries, is now becoming sufficiently reliable so that we can exploit this source of genetic material via *in vitro* fertilization and embryo transfer. Because costs per animal for embryo transfer remain high, most applications will be in cattle for the next decade (a superior calf is worth much more than a superior lamb). Two properties of harvesting female gametes in this way are noteworthy. The first is reproduction from a dead animal. Thus a valuable cow or horse that dies or is slaughtered due to terminal illness can reproduce; note that this procedure works at any stage of the reproductive cycle and even during pregnancy. Another advantage is to exploit genes of slaughtered females with especially good carcasses such as high dressing percentage, tender meat with minimal fat, etc.

The second noteworthy property of embryos produced in this way is that they can be produced in huge quantities at low cost. This makes them appropriate for applications such as twinning cattle, and for other biotechnologies like cloning by nuclear transplantation. This also makes them ideal for many experimental purposes. A special fringe benefit is that experiments can be done without using experimental animals, just nearly worthless ovaries from commercially slaughtered cows plus frozen semen.

In vitro fertilization and culture of embryos are required to exploit many applications of *in vitro* oocyte maturation, but there are some specific uses of these technologies in addition. For example, *in vitro* fertilization can be used with technologies of sexing semen that produce too few sperm for artificial insemination but plenty for *in vitro* fertilization. *In vitro* fertilization also can be used for testing fertility of males, which is extremely costly and frequently inaccurate with other procedures.

The second from last item in Table 3 concerns melatonin. Sheep and goats are seasonal breeders, which causes great constraints in agricultural productivity. They normally become pregnant when days become short and the nights, long. The pineal gland at the base of the brain secretes melatonin when it is dark and fails to do so in daylight; moreover, melatonin is a key regulator of reproductive hormones in these species. Therefore, one can cause the same effects as darkness by feeding melatonin. A feeding at 4 p.m. is the equivalent of darkness commencing at that time. After a few weeks of this animals resume fertile reproductive cycles and if mated, become pregnant, no matter what the season of the year.

The last item in Table 3, twinning in cattle, can be induced in several ways, including embryo transfer or injections to increase double ovulations. These procedures are just becoming economically feasible because of combining some of the technologies just described. The main principle that makes twinning attractive is that about 70 percent of the nutrients that a beef cow eats in the course of a year go for her maintenance requirements, and the remaining 30 percent go to needs of pregnancy and lactation. Thus for 30 percent more feed, one can produce two calves instead of one. There is, however, higher morbidity and mortality with twins, and more labor is required.

PROMISING BIOTECHNOLOGIES NOT YET COMMERCIALIZED

In addition to standard recombinant DNA procedures, there are numerous extremely powerful biotechnologies being used in the laboratory that

...although transgenesis is one of the most powerful technologies available for scientific endeavors, it is not likely to have much impact in production agriculture over the next decade.

will eventually impact on animal agriculture. A somewhat arbitrary list is in Table 4. A few of these items will be discussed further. The first, transgenic procedures, is covered in more detail elsewhere in these proceedings. I want to emphasize that although transgenesis is one of the most powerful technologies available for scientific endeavors, it is not likely to have much impact in production agriculture over the next decade. By far the main problem is that we know so little about gene function for production traits in farm animals that we simply do not know what genes to manipulate, much less how to manipulate them. This clearly will change eventually, and although there may be several spectacular applications before the year 2000, most will occur after that date. Other problems with transgenic technology include expense and the extremely long time-frame involved with species with long generation intervals. Sexing semen is another promising biotechnology. At least one method of sexing semen, that of using a flow cytometer, has been convincingly demonstrated to sort X and Y chromosome-bearing sperm of several species with 80–90 percent accuracy. While the sperm are damaged somewhat, and the procedure is much too slow to be used for routine artificial insemination, the sperm are reasonably fertile and could be used for *in vitro* fertilization. There is a huge and immediate market for sexed semen in cattle.

TABLE 4 EXAMPLES OF FINDINGS OR BIOTECHNOLOGIES THAT LIKELY WILL HAVE APPLICATION IN PRODUCTION AGRICULTURE IN THE FUTURE

Transgenic procedures
Polymerase chain reaction
Sexing semen
Embryonic stem cells
Homologous recombination
Artificial chromosomes
Somatotropin and beta agonists for meat production
Growth factors
Second messenger systems
Transcription factors and other regulators of gene expression
Early pregnancy factor
Trophoblast-specific interferons
Marker-assisted selection

GROWTH FACTORS, TRANSCRIPTION FACTORS, AND SECOND MESSENGER SYSTEMS

These fall into the category of important areas of research concerning regulation of cellular function; thousands of research papers are published in these areas annually, and the resulting information will be exceedingly useful in getting cells to do useful things. A decade from now, some of these will reach the interface with production agriculture with very important consequences. However, much more information is needed to make educated guesses concerning exactly which factors will be matched with which applications.

New biotechnologies as well as new applications of older biotechnologies continue to be applied to increase the efficiency of animal agriculture.

Extolling the virtues of marker-assisted selection is currently very much in fashion. Related procedures using restriction fragment length polymorphisms (RFLP) form much of the basis for projects on mapping (not sequencing) human and other genomes. Similar procedures have been successfully applied to finding genes responsible for several human diseases. An example of how this concept works is that one tries to match up specific and easily measured genetic differences (e.g., a particular RFLP) with a desirable characteristic (e.g., a high percentage of unsaturated fat in meat) so as to quickly and easily identify the genetically superior animals in the population. A fringe benefit is that one can further exploit such matchups to locate and identify the genes responsible for the desired characteristic.

This important biotechnology has many potential uses. Unfortunately, there are also severe limitations because dozens of genes are responsible for most traits of interest (e.g., fertility, growth rate, disease resistance), which frequently results in confusion with this approach. However, in other instances, especially where single genes with multiple alleles have huge effects, this approach is excellent.

BIOTECHNOLOGY NEEDS IN ANIMAL AGRICULTURE

If one asked about biotechnological needs, one likely would get quite different answers from different groups, e.g., farmers vs animal scientists. I have listed some of the more obvious ones in Table 5. Many of these have been alluded to earlier; the usefulness of most is obvious. A number of those listed would result in markets in excess of $100,000,000 annually. Some would lead to entirely new approaches and greatly increase efficiency, e.g., cloning animals from adult cells.

All of these organisms are more complicated than super computers or space ships.

TABLE 5 BIOTECHNOLOGICAL NEEDS IN ANIMAL AGRICULTURE

Gene maps
Basic information about all aspects of animal biology, especially appetite, stress, disease
Sexed semen (eggs for poultry)
Improved techniques to crypreserve poultry semen
Cloning animals from adult cells
Reliable and simple transgenic technology
Early pregnancy tests
In vitro gametogenesis
Methods of modifying animal products for fat and other characteristics
Quality control for salmonella, etc.
Very inexpensive diagnostics for on-farm use
Immunocastration
Reprogramming laying hens

BOTTLENECKS TO APPLYING BIOTECHNOLOGY TO ANIMAL AGRICULTURE

Some obvious bottlenecks are listed in Table 6. Interestingly, they fall into two broad categories: biological constraints and sociological/societal constraints. Farm animals are much more complex than single cell organisms. All of these organisms are more complicated than super computers or space ships. Therefore it takes enormous resources to understand and manipulate them to useful ends, including modifying organisms genetically. However, there is considerable optimism in scientific circles in dealing with this obstacle. It is considered challenging, interesting, and important, and progress is being made. The complexity of cells and organisms is truly amazing. Even non-cellular entities such as viruses are complex, and their interaction with cells especially complex. A graphic example is the human immunodeficiency (AIDS) virus; direct and indirect funding to understand this virus is approaching one billion dollars annually on a world-wide basis, and while information is accumulating exponentially, many questions remain concerning effective means of controlling this virus other than by human behavioral modifications. Note also that one billion dollars is really minuscule compared to military expenditures.

The other items in Table 6 are also bottlenecks, some of them diffuse ones. Any one or two of these might be dealt with easily, but their cumulative effect can be devastating. For example, fewer students are undertaking

TABLE 6 BOTTLENECKS TO APPLYING BIOTECHNOLOGY TO ANIMAL AGRICULTURE

Complexity of organisms
Diffuse, underfunded efforts
Instability of funding
Secrecy due to proprietary considerations with many sources of funding
Research climate that discourages young scientists
Costs of FDA approval process
Regulatory costs of doing research
Time required to prepare proposals, get committee approvals, prepare progress reports, etc.
Narrow training and experience of most life scientists
Lack of clear goals
Remoteness of many scientists from needs of farmers

graduate studies in animal biology than in the past, which perhaps is desirable since employment opportunities are limited. Another very interesting phenomenon is the change from predominantly men to predominantly women graduate students in these areas. It appears that fewer young men find this area of endeavor attractive in the United States, at least in part due to some of the items in Table 6.

Despite the cumulative insidiousness of the items in Table 6, many are desirable and necessary, and most alternatives to them are even less desirable. Thus, the solution is to deal with the items in Table 6, even though this takes huge amounts of time and funding.

SUMMARY

Hundreds of biotechnologies are applied daily in animal agriculture in the United States. New biotechnologies as well as new applications of older biotechnologies continue to be applied to increase the efficiency of animal agriculture. At the same time, new and more varied animal products are being produced. Because there are considerable time lags between conceiving a new application and its use in production agriculture, it is too early to evaluate eventual impacts of newer technologies such as recombinant DNA procedures. However, it is likely that they will be exceedingly important for animal agriculture in the decades ahead. Although there are considerable regulatory and safety costs in applying biotechnology, by and large, these costs are reasonable and need to be taken into account in planning and budgeting.

SELECTED REFERENCES

Evans, J.W. and A.Hollaender, eds. 1986. *Genetic Engineering of Animals: An Agricultural Perspective.* 328 pp. Plenum, New York, NY.

Hansel, W. and B.J. Weir, eds. 1990. Genetic Engineering of Animals. 240 pp. *J. Reprod. Fertil. Suppl.* No. 41.

Johnson, L.A., Flook, J.P. and H.W. Hawk. 1989. Sex preselection in rabbits: live births from X and Y sperm separated by DNA and cell sorting. *Biol. Reprod.* 41:199–203.

Juskevich, J.C. and C.G. Guyer. 1990. Bovine growth hormone: human food safety evaluation. *Science* 249:875–884.

Seidel, G.E. Jr., ed. 1985. *Technology in Animal Agriculture: A Seminar for Investment Strategists.* 337 pp. Colorado State University, Fort Collins, CO.

Seidel, G.E., Jr. and R.P. Elsden. 1989. *Embryo Technology.* 101 pp. W.D. Hoard & Sons, Fort Atkinson, WI.

Institutional Incentives & Impediments

BIOTECHNOLOGY AT THE CROSSROADS
IS REGULATION THE GATEKEEPER?

ROBERT B. NICHOLAS

Partner
McDermott, Will & Emery
1850 K Street, NW Suite 500
Washington, DC 20006-2296

Nicholas heads the firm's Biotechnology Practice Group and specializes in agricultural and environmental applications of biotechnology. He is also Executive Director, NABC.

consumer acceptance and scientific problems...will continue to slow the pace of approval...

As of May 1991, the United States Department of Agriculture (USDA) had approved approximately 120 applications for small-scale field trials of transgenic plants and several genetically engineered animal biologics had been approved for marketing. The Food and Drug Administration (FDA) had approved for commercial use one biotechnologically derived food enzyme and the Environmental Protection Agency (EPA) had allowed a handful of genetically engineered pesticides and other organisms to be used in the field and in closed systems. As the science of molecular biology appear to progress, quickly bringing agricultural applications of biotechnology closer to the marketplace, we, the organizers of this third NABC meeting, thought the time appropriate to focus our discussions on biological, social and institutional questions voiced about biotechnology and how concerns might affect the pace of development and commercialization of agricultural biotechnology.

My task today is to review the regulatory structure for the agricultural products of biotechnology. After a brief introduction, where I adopt the viewpoint that agricultural biotechnology is at "a crossroad," I will present, in an historical context, an overview of the current regulatory system and conclude with an evaluation of the effect of the regulatory framework on the pace of commercialization. My conclusion is that consumer acceptance and scientific problems, as well as problems in the regulatory system, will continue to slow the pace of approval of the products of agricultural biotechnology.

IS AGRICULTURAL BIOTECHNOLOGY AT A CROSSROADS?

There is little doubt that agriculture has been, and regardless of the pace of regulatory approvals, will continue to be greatly affected by biotechnology. New crops are being developed, as are new animal drugs and vaccines, and there is an explosion of understanding about the basic biology in agricultural systems. Today there are several new agricultural biotechnology companies and most of the large traditional players in the agricultural sector—seed companies, pesticide companies, livestock producers and food processors—have biotechnology initiatives. Therefore, from a long-term perspective, agriculture and biotechnology appear as healthy partners.

From a short-term commercial perspective, however, agricultural applications of biotechnology appear to be at a crossroads. There are many strong supporters for the majority of agricultural applications of biotechnology in agriculture. However, unlike their biomedical cousins, which have been viewed from the outset as valuable and have been commercialized almost without controversy, questions have been raised by some from the very beginning about the need for, and the social, political, economic and safety implications of initial targets for commercialization of biotechnology in agriculture. Wide-ranging public debate has taken place long before any products have entered the commercial arena. Focusing not just on the traditional concerns of environmental and human safety, but on new criteria such as potential economic, social and philosophical issues, the debate has made the process of formulating a rational regulatory scheme more uncertain and protracted. Coupled with a shortage of agency resources, the pace of agency consideration and development of policies toward products of biotechnology used in agriculture has been slow and the resulting uncertainty and cost has caused a number of the smaller companies to depart from the industry as well as a reduction in the efforts of some larger companies.

Therefore there is a "crossroad"—the smaller companies may not have the wherewithal to last through the continuing debate and regulatory uncertainty. In my view, the question is not whether biotechnology will be commercialized in agriculture, but when and by whom. Of the four determinants—science, intellectual property, regulations and acceptance by consumers, farmers and processors—I will be focusing on the latter two. The question I ask at this point is, "Have we created a system in which only the biggest or largest can survive?"

OVERVIEW OF THE REGULATORY FRAMEWORK

In June 1986, the President's Office of Science Technology Policy (OSTP) published the Coordinated Framework for the Regulation of Biotechnology (51 *Federal Register*, 23302 *et seq.*, June 26, 1986). In 1987, USDA published a revision to the 1986 proposal, (52 *Federal Register* 22892, June 16, 1987). These documents contain the basic framework for the regulatory oversight of the products of biotechnology. The development of the framework can be traced back to the 1970s.

Have we created a system in which only the biggest or largest can survive?

HISTORICAL LESSONS

The history of recombinant DNA regulation, a period of only 15 years, is instructive for the current debate. The history can be viewed in three parts. The first period, ending in about 1978, was characterized by conjectural fears about laboratory-based research, raised initially by the scientific practitioners and then spread into the political and public arena. Notwithstanding numerous Congressional hearings and media attention, the issues were largely laid to rest as a result of the adoption of "guidelines" issued by the National Institutes of Health (NIH). The NIH Guidelines set the standard and today, as laboratory experience has allowed relaxation of the Guidelines, 90 percent of laboratory research is exempt from the Guidelines.

Some observations about this period are instructive when contrasted to the current debate. One, a scientific and political consensus was developed within a relatively short period of time, approximately three years. The NIH played a major role, together with the scientific community, in helping to allay public fears and put in place flexible, yet credible, guidelines. Two, the regulations were entirely process-based; that is, the regulations focused entirely on recombinant DNA molecules; there was no efforts to recast all regulatory authorities to the new science. Three, the Guidelines focus on lab experimentation, primarily in the biomedical field; environmental releases were generally prohibited.

The second period of the regulation, or oversight of recombinant DNA technology, begins in the early 1980s and was highlighted by the commercialization of the initial biomedical applications of the technology and the development of potential agricultural and environmental applications of biotechnology. The biomedical field turned out to be largely non-controversial; agricultural applications sparked a significant debate which is on-going.

Two observations seem appropriate. First, because the initial biomedical products were viewed as potentially lifesaving or dealing with serious diseases, none raised significant public policy questions. Second, the products were the providence of one agency—FDA—and traditional concerns were focused on in the FDA regulatory structure, e.g., safety and efficacy[1].

In contrast, agricultural and environmental biotechnology products were viewed as presenting a very different perspective because living organisms were to be introduced into the environment or intended for human consumption. Started by a series of congressional hearings before the Oversight Subcommittee of the House Science and Technology Committee, and continued with a review by the White House, the new applications of biotechnology were examined to determine whether the current laws and regulations were adequate to protect the public while not unduly impeding the technology. It is this debate that is still going on, now almost a decade later.

As before, several observations seem useful. First, the focus of the debate shifted from laboratory experiments by academic researchers and development of lifesaving medicines, to planned introduction by companies into the environment of genetically modified plants, animals and microbes and to introduction into the food chain of food products from these organisms. Second, because of the wide ranging potential uses of these new products several agencies were now involved, frequently in areas where they either had not previously regulated or in overlapping areas. In addition to FDA, agricultural applications involve principally EPA and USDA. As a result, there have been and continue to be, disagreements

---

[1]While FDA has had to address many issues involved in evaluating new biotechnology products, the issues have been mostly traditional concerns in the drug approval process.

about the scope of the organisms reviewed, agency jurisdiction and risks and benefits of specific applications.

The third period of regulatory oversight of biotechnology begins with the publication of the Framework in 1986 and continues into the present. During this period the agencies reviewed various applications involving mostly small scale field trials. While unsuccessful, there were also many EPA efforts to modify the Framework in light of increasing experience and efforts by USDA to develop guidance for large scale testing and commercialization of biotechnology products.

The Framework addresses the jurisdictional basis of the regulation of the products of biotechnology; it describes EPA's, USDA's and FDA's jurisdiction, together with that of NIH and the Occupational Safety and Health Administration. The Framework is grounded in four principles. The first is that existing statues are sufficient; *e.g.*, no new laws are necessary. The second principle is that the product, and not the process used to develop the product, would be the regulatory focus—the basis of regulation would be the end product and not how that product was produced. The third is that regulation would proceed on a case-by-case basis, and not by class, until there was sufficient information to make determinations about the safety or lack of safety of classes of products. The fourth principle is that the agencies would attempt to coordinate and to work reviews in conjunction with one another, rather than sequentially.

The Framework principally focuses on small-scale introductions and, while describing in some detail agency programs for approval of traditional products, it provides little guidance on large scale and commercial issues. Several applications, for example, transgenic animals, are not discussed, nor are regulatory questions with respect to commercialization of products of transgenic plants, the registration of pesticides or herbicide-tolerant crops Notwithstanding its deficiencies, the agencies and particularly USDA, have used the Framework as a mechanism to allow some research applications to move into the field trial stage.

A brief description of each agency's regulatory program follows.

USDA—The Animal, Plant Health and Inspection Service (APHIS), relying on the Federal Plant Pest Act (7 U.S.C. §§ 150aa–150jj) requires that a permit application be filed before movement, defined to include an environmental introduction, of any organism produced via genetic engineering which has a plant pest source (donor, vector or recipient) and any organism having a potential plant pest risk. Principal concerns of APHIS have been the source of the organism and its ability to present a risk to agriculture. The agency has also used its authority under the National Environmental Policy Act (42 U.S.C. §§ 4321–4370) to conduct environmental assessments.

Principal concerns of APHIS have been the source of the organism and its ability to present a risk to agriculture.

To date, APHIS has granted approximately 1,000 permits for movement of genetically engineered organisms for lab research and approximately 120 permits for actual field trials. Of the field trial permits, about 40 percent have involved herbicide tolerant plants, with the balance for pest and viral resistance and some few for plants with compositional modifications. In addition to tobacco and tomatoes, the initial targets, many vegetable and some tree and fruit crops, have been tested in the field.

Under the authority of the Virus, Serum and Toxin Act (21 U.S.C. §§ 151–158) APHIS regulates production and sale of animal biologics, such as animal vaccines and some diagnostic tests used in the treatment, prevention, cure and diagnosis of diseases of animals. APHIS has approved three licenses for a recombinant vaccine and approved field testing of another vaccine. The basic questions asked by APHIS in the review are traditional to agency concerns: potency, purity, efficacy and safety. APHIS also evaluates environmental considerations and on one such occasion, environmental issues have held up a permit approval.

Most observers believe that APHIS's program for review and permitting small scale trials of plants has been successful; APHIS has allowed for public and state participation and has conducted credible reviews on a timely basis. APHIS also has sought to stimulate thinking about downstream considerations by sponsorship of conferences, such as the meeting on transgenic maize held at the Keystone Center.

EPA—EPA's biotechnology policy relies on its authority under the Federal Insecticide, Fungicide and Rodenticide Act (FIFRA) (7 U.S.C. §§ 136–136y) and the Toxic Substances Control Act (TSCA) (15 U.S.C. §§ 2601–2629), to regulate pesticides and chemicals respectively.

Chemicals, which EPA has defined to include DNA, are subject to a 90-day premanufacture notification requirement, if they are not on EPA's inventory of existing chemicals in commerce. EPA's review has focused on enhancement of the host range of the modified organism and the ability of the organism to affect the ecosystem and human health. EPA has received notification for small-scale field testing of microbes used for nitrogen fixation and for several genetically modified microorganisms used in closed systems to produce enzymes for non-FDA regulated uses. (The enzyme itself is treated as traditional chemicals for regulatory purposes.) Taxonomy has been a major issue for EPA as well as definitions for the scope of organisms subject to review.

Under its FIFRA policy EPA focuses on the small scale field testing of biological pest control agents. EPA adopted a policy of requiring agency notification prior to any field testing of certain genetically modified organisms. EPA has received several notifications, and most small-scale trials have proceeded without the need for an experimental use permit. (Field tests of chemical pesticides on less than ten acres remain exempt from notification and experimental use permit requirements.)

EPA has not been successful in modifying its 1986 policy statement notwithstanding several efforts that have come to the proposal stage. Disagreements about the scope of organisms subject to notification and regulation, as well as agency turf battling, have led to this stalemate, but many are hopeful that the current proposed revisions will be published and the public debate will begin. Not addressed in the revisions, and still to be decided are the downstream issues: how to regulate, if at all, plants with increased pesticidal properties, whether to regulate herbicide tolerant crops, and how to address the tolerance/residue issues.

FDA—FDA regulates in several areas which affect the agricultural applications of biotechnology including food and animal feed, food additives, animal drugs and possibly transgenic animals. FDA's basic authority is the Federal Food Drug and Cosmetic Act (FFDC Act) (21 U.S.C. §§ 301–392). While FDA has approved several genetically engineered foods, animal biological for human use, no genetically engineered foods, animal drugs or feed additives have been approved as yet. One food enzyme has been approved.

FDA has not generally provided guidance as to how it will regulate the products of biotechnology, though it has stated on more than one occasion that it will rely on its traditional programs. FDA does not regulate foods on a pre-market clearance basis. Traditionally bred crops have been marketed without FDA clearance, subject only to the general responsibility of the food company to assure the safety of the food, e.g., conformance with the adulteration provision of the FFDC Act. Food additives or substances which are added to foods or become constituents of foods, must be "generally recognized as safe" (GRAS), or the subject of an approved food/feed additive regulation.

Animal drugs are also regulated by FDA in much the same manner as human drugs. In addition to proving target animal safety and efficacy a sponsor must also demonstrate that the drug does not produce harmful residues in edible food. An additional difficulty for those involved in the animal health field is that there is a split of jurisdiction between FDA and USDA, with FDA regulating animal drugs and some diagnostics and USDA regulating animal biologics and some other diagnostics. How ones' product is categorized has significant ramifications, with respect to which agency regulates, as well as to how the product is regulated.

THE REGULATORY ROADBLOCK?

While the Framework has permitted significant progress since its adoption in 1986, major uncertainties exist with regard to the future. These uncertainties will need to be resolved before one working in the field can have any reasonable degree of assurance that a safe product will be allowed to reach the marketplace on a timely basis and become a success or failure, depending on its value to the consumer. In my view, three things must happen so that the regulatory process does not become a roadblock.

The first is establishment of clear jurisdictional lines. While many of the questions as to "who" regulates the "what" have been decided with respect to field testing, significant questions remain unanswered regarding large scale testing and commercial approval. For example, it is unclear who will regulate commercialization of transgenic crops with enhanced pesticidal properties and how food and feed tolerances will be set. No clear policy has been articulated for transgenic animals. Questions also remain about the scope of the organisms subject to initial or commercial oversight.

A regulatory roadmap would give the regulated industry, the scientific community, public interest groups and consumers answers to questions—who has jurisdiction, how that jurisdiction will be exercised, and what is expected in order to obtain regulatory decisions.

Second is credibility. The agency charged with the responsibility must be able to render a credible decision. As an initial matter, the agency must have the necessary expertise. The regulatory process also must allow for appropriate public participation of decisions and communication of results.

A significant question yet to be answered is whether the regulatory process will focus on safety and efficacy, or whether the regulatory process will become the place for protracted debates about potential social and economic implications of the technology. This question is an essential one for the agricultural sector where the primary debate about products like bovine somatotropin have very little to do with safety and more to do with potential questions about the changing structure of agriculture.

A significant question...is whether the regulatory process will focus on safety and efficacy, or whether the regulatory process will become the place for protracted debates....

Third is timeliness; can the agency make the decision on a timely basis? As an initial matter, the agency must have sufficient resources. The agency's process must also provide reasonable time frames for decisions. A decision that takes an unnecessary amount of time, particularly in an area where the technology is moving rapidly, benefits no one.

CONCLUSION

As I have discussed, I do not believe that regulatory processes will be a roadblock to agricultural biotechnology in the long term. However, in the short term, the pace of approval of the products of agricultural biotechnology will be slowed by uncertainty.

NEEDED REFORMS IN THE HARMONIZATION OF U.S. PATENT LAW

WLLIAM LESSER

Agricultural Economics
Cornell University
Ithaca, NY 14853

A Professor of Agricultural Economics at Cornell University, Lesser also directs the Western Societies Program. His research interests encompass livestock marketing, assessment of the impacts of biotechnology products and the implications of patents and other forms of intellectual property on agriculture worldwide.

It is a great pleasure to have this opportunity to discuss the significant area of intellectual property. The time allotted to me is brief so my remarks must be limited. What I shall attempt to do is to identify major areas where, in my judgment, intellectual property laws need change, or at least attention. Please remember that I speak as an economist, not as a lawyer, and my comments will reflect that focus on the general economic role of intellectual property, not on the particulars of any case. Specifically, emphasis shall be on the role of intellectual property in the stimulation and commercialization of agricultural biotechnology products and changes which would (could) enhance that stimulative effect.

In particular, I shall comment on the following topics:

—imports of "products-by-process,"
—broad protection for pioneering inventions and the rush to invent,
—the U.S. first-to-invent system and international harmonization,
—clarification of the experimental exception from infringement, and
—implications of revisions to the International Union for the Protection of New Varieties of Plants (UPOV).

Those who follow these matters will recognize that my comments are directed principally to patents and related forms of intellectual property rights, not to copyrights and trademarks. That seems appropriate as patents are the most suitable form of protection for the types of products represented at this meeting. Before discussing patents in some detail, it may be helpful for those in the audience whose only direct contact with patents is the notice on a can opener reading "patent pending" (incidentally a legal status in the U.S. which has no real significance whatever), to give some broad overview of what intellectual property is and how the laws function.

Intellectual property laws are a form of economic incentive which operate through the ability to exclude others.

INTELLECTUAL PROPERTY

Intellectual property refers to creations of the mind—inventions, music, books, etc. Such things require special attention because, unlike other more tangible forms of property, they have a "common good" attribute—"consumption" by another person does not directly diminish the amount available. We can all sing a hit song without its availability to anyone being reduced. Clearly we cannot all drive the same car or live in the same house without someone being disadvantaged.

Common property is one of the theoretical (and practical) justifications for government involvement. National defense is a common property good provided by governments and so is agricultural research like plant breeding which worldwide is and remains largely government funded[1]. The private sector can involve itself, but the catch is no profits can be claimed. If everyone can sing my song free of charge, I would have little economic incentive to be a composer/lyricist. At best I would be a part time one, working the remaining time to support my song writing "hobby."

At this point, intellectual property comes in, providing an incentive for private entities to engage in creative activities. This is done by providing the creator/inventor, the holder of the right, with the option of prohibiting others from using his/her creation. In parallel with the right we have of excluding others from using our private real property. From this perspective two major conclusions can be drawn about intellectual property:

—its intent is economic, the provision of a monetary incentive to attract private investment to creative endeavors,[2] and

—it works by exclusion, the ability to prohibit others from using one's invention.

Intellectual property law then allows a limited (in scope), temporary (for a specified period) monopoly right over inventions. Of course, there is little

[1] For a further explanation of these issues see any text on public finance economics, eg., Herber, 1975.

[2] Another approach to intellectual property is that of inherent rights–an inventor should have the exclusive right to use his or her invention. That approach is not treated here in favor of the economic incentive concept.

This all says intellectual property is less than many assume it to be.

real benefit in simply excluding others, so permission is generally given for compensation—the royalty payment.

Intellectual property laws then are a form of economic incentive which operate through the ability to exclude others. The legislation says nothing about one's freedom to use the invention. That right might be limited by regulation (as described in the preceding paper by Robert B. Nicholas), by a related, prior grant of protection, or by other means. Nor does the grant of protection say anything about the practical significance of the invention/creation. It is not, nor could it be, the responsibility of government employees in, say, the Patent Office to predict the economic viability of an invention. That decision is made by the market. The inventor has the opportunity to profit from the invention, but no guarantee. Indeed, only something like 20–30 percent of patents are ever commercialized and few of these are really lucrative. Moreover, it is the responsibility of the holder of the right to enforce it. We do not have a police force for infringers; rights must be protected through court proceedings initiated by the involved parties.

This all says intellectual property is less than many assume it to be. It is important, but generally not that important. Referring again to patents, businesses often rank patents low on lists of decisions on where/how to invest in research and development (surveys in Scherer, 1980, p. 446; Nogués, 1990, pp. 11–14). But note that where it is ranked high is for pharmaceutical products and living organisms, which are easily copied. As these products are the underpinning of much of the biotechnology industry, that sector is especially concerned with the form and operation of intellectual property protection.

Since intellectual property laws serve as an economic policy tool, a device to spur private research and development spending, it would be satisfying to know how well this policy works. Regrettably that question cannot be answered with any real clarity. We have some evidence that such laws do spur investment (Butler & Marion, 1983). What is lacking is evidence that existing laws are anywhere near optimal in terms of scope or length of protection (see the literature review by Braga in Siebeck, 1990). This is an important issue because intellectual property laws consist of many detailed aspects for which we have little economic evidence to guide us. Many of my comments below deal with such aspects, but recognize there is no firm evidence to point to in support of one position or the other.

...the U.S. has the broadest protection available in the world.

Since intellectual property comes in numerous forms—music, inventions, books, computer programs, data bases, etc.—several laws are required to cover this breadth. The major of these, and the principal creative areas they apply to, are as follows, recognizing that any terse list like this does grave injustice to the nuances of these issues:

Patents: inventions for products and processes (a special form for selected plants is known as Plant Breeders' Rights);

Copyright: books, music, recordings, computer programs;

Trademark: product names and other identifiers;

Trade secrets: anything of economic value which is actively kept from the public.

The first three of these have their separate requirements and functions, but those details exceed our scope at this point. The fourth category, trade secrets, acts fundamentally differently, for it relies on secrecy rather than disclosure. It is also perpetual, so long as the secret can be maintained (as with the formula for Coca Cola syrup). But perhaps most significantly trade secrets in the U.S. are based on state rather than national law and hence protection is more variable. Contracts often substantiate the basis for this protection or extend it into other realms.

When considering the overall protection allowed by intellectual property laws it is important to recognize that they are often used jointly. For example, technologies can be licensed with a patent licensing agreement for the base technology accompanied by trade secrets (possibly involving non-disclosure agreement) providing some "tricks of the trade" of the most efficient use of that technology. Thus intellectual property protection involves a range of types of laws. Adding to that the difficulty of enforcement in some instances, the determination of the actual degree of protection for a particular product/technology in a specific country is indeed a complex matter.

That said, the U.S. has the broadest protection available in the world. The breadth applies particularly to higher plants and animals where patenting is a matter of policy and practice. Contrast that with the European Community which is still struggling with the patentability of these products (see the OECD Directive, 1988) although progress is now being made.

...limitations to intellectual property protection in the U.S. have not been a major factor in delaying private research and development investment in biotechnology.

In fact let me make the broad statement, without much specific substantiation, that limitations to intellectual property protection in the U.S. has not been a major factor in delaying private research and development investment here in biotechnology. Those who are looking for a major source of the problem will have to look elsewhere, in my judgment. This of course does not say everything is fine in this area. Let us now look at where improvements seem to be needed.

IMPORTS OF PRODUCTS–BY–PROCESS

For many years, U.S. inventors lacked effective protection for products produced overseas by a technology patented in the U.S. The rights granted read as follows (35 U.S.C. Sec. 271—Infringement of Patent): "...Whoever without authority makes, uses or sells any patented invention, within the United States during the term of the patent...." Clearly, as the use of the technology did not occur in the U.S. there is no violation of this statute. Contrast this position with the rights specifically granted in many countries. The European Patent Convention for example reads in Article 64 (2), "If the subject-matter of the European patent is a process, the protection enforced by the patent shall extend to the products directly obtained by such a process." Protection for U.S. inventors, however, exists outside the patent act—specifically in the 1988 Omnibus Trade and Competitiveness Act (Biotechnology Newsletter, Nov. 1988).[3] That Act purposely avoids the term "directly obtained", substituting two provisions delineating when infringement has not occurred:

1–[the product] is materially changed by a subsequent process, or
2–[the product] becomes a trivial and nonessential component of another product.

While these matters may seem esoteric, they have a direct bearing on agrobiotechnology. For example, as Bent, *et al.* (1987, pp. 320-22) discuss, how are the products of a potential breeding process or gene insertion system to be interpreted? The first generation (the direct product) using these technologies would almost certainly be covered. But what of succeeding products or generations which are the creation of the patented technology

[3]An earlier version, the Tariff Act of 1930 in Section 337 did grant similar protection, but its breadth raises problems, especially with GATT (see Barton, 1989).

but did not directly employ it? In the U.S. this matter can be partially avoided by claiming the propagation method, but that necessitates identifying all the possible products of the invention. In Europe the ban on the protection of the products of "essentially biological processes" (EPC Article 53 (b)) prevents that approach.

The matter is a complex one for limiting protection to only the direct products likely gives too limited protection. The extension to all future generations, however, gives a potentially enormous scope, something that clouds the ownership of many living organisms. That extent of protection seems excessive and will in some cases chill further research. The normal pattern is for case law to resolve these matters, but that is a slow and imprecise method. It is better to write the legislation more specifically to identify, for example, how many generations are to be covered in a product-by-process claim. My preference for the U.S. is to place such legislation within the Patent Act.

However, the purpose of intellectual property is not equity to the creator, it is the social good to which these laws are directed.

BROAD PATENT PROTECTION FOR PIONEERING INVENTIONS

If my preceding comment could be broadly supported, this one will be far more controversial. It has to do with the essentially "defensive" position from which U.S. patent examiners operate. In brief, they must grant the claims made by the inventor unless specific reasons (evidence) for denying them can be established. The evidence required is prior knowledge or prior inventions. Inventors in new, pioneering endeavors will typically claim broadly, and the very newness of the field means examiners have no basis for denying those claims. A key example is that of the "Harvard mouse" — the one and only higher animal patent to date—which claims "A[ny] transgenic non-human mammal. . . ." (claim #1, U.S. Patent No. 4,736,866).

At one level this sort of broad claim granting seems equitable—something pioneering, after all,, deserves a greater reward. However, the purpose of intellectual property is not equity to the creator, it is the social good to which these laws are directed. Society benefits by bringing forth an invention at the least cost, that is, by granting the smallest degree of monopoly rights needed to induce an invention. From what we know about fundamental inventions, they are inspired by some form of creative drive, not solely by financial motives. Patents seem best suited to stimulate

Narrower but clearer protection would seem to be preferable in business planning to broad but fuzzy protection.

the repetitive work required to refine inventions and make small ongoing improvements (Scherer, 1980; Jewkes, *et al.*, 1969). Therefore it is not clear if society is receiving a proper return for the broad protection granted.

This matter touches on two issues of patent law and patent economies known as the rush to invent and undue experimentation. The rush-to-invent analysis recognizes that the patent system grants great benefits to the successful—the first—and little to others (but see comments under the first-to-invent system below). This in turn can stimulate hurried, duplicated research by multiple firms, a social waste. Secondly, firms fearing they are late entrants may desist from entering altogether. Counterbalancing these, the race tends to bring forth inventions more rapidly than otherwise, a societal gain. It has not been possible to determine theoretically or empirically where the balance lies (Siebeck, 1990; Scherer 1980, Chap. 15; Dasgupta and Stiglitz, 1980). The granting of large value to pioneering inventions however exacerbates these issues although again the societal outcome is not clear.

"Undue experimentation" is a patent term which says in essence that the invention must be replicable in some reasonable time or economic frame (see Van Horn, 1987). It provides a way that examiners can limit a patent scope by requiring proof of duplicability. This issue for example might arise in the "Harvard Mouse" patent where all non-human mammals are claimed, but to date (to my knowledge) only a mouse has actually been produced. Can indeed the procedure be extended to primates without undue experimentation? This matter will likely be reviewed in appeal or in the courts in the future.

The U.S. is nearly alone in granting patents to the first-to-invent.

If the theoretical issue is not clear, the practical considerations are becoming so. Without attempting to comment on the merits of any particular patent, broad legal battles are emerging now in the area of agricultural biotechnology. This is particularly true for broad claims to a method of achieving some goal, that is to say to a claim to the dominance of a technology or procedure. It is not clear, at least to me, that society benefits from such broad grants. What it may lose are the costs of the litigation along with the chilling effect such unclear rights have on related research. Narrower but clearer protection would seem to be preferable in business planning to broad but fuzzy protection. I call for a reform in Patent and Trademark Office practices in this area.

THE U.S. FIRST–TO–INVENT SYSTEM

One of the steps in awarding a patent is determining who is the inventor. Most of the world follows a simple system—the first to file is presumed to be the inventor (the first-to-file system). The U.S. is nearly alone in granting patents to the first-to-invent– an extreme that allows a transfer of the patent even after it has been granted. The process followed is one of interferences (35 U.S.C. Sec. 135). Considerable documentation, including detailed, dated lab roles, is required to substantiate the date of invention.

Conceptually and theoretically speaking, the first-to-invent system is preferable. It awards the patent to the true inventor and reduces some of the pressure for the race to invent, mentioned above. Moreover, the first-to-file system combined with a grace period (the time prior to the first patent application) allows some opportunity for a strategic revealing of the invention to foreclose rights to others (Lesser, 1987).

That said, the first-to-invent system, too, raises problems, the greatest of which is protracted court battles. One was recently concluded in favor of Amgen over ownership of the patent for recombinant erythropoietin (EPO) (Nature 350 (1991):99; Biotechnology 9(1991):327). Moreover the documentation process is unfair to foreign inventors as certain lab documents must be certified in the U.S. and burdensome to small inventors. It is, in my judgment, time for the U.S. to abandon our first-to-invent system and join the rest of the world.

CLARIFICATION OF THE EXPERIMENTAL EXEMPTION

It is not necessary to convince agricultural researchers of the need for access to prior developments—patented or not—as part of the continuum of developmental work. To this group the words of the Patent Act (35 U.S.C. sec. 271) "...whoever without authority makes, uses or sells..." is indeed chilling. What will their access be to patented plants and animals? This issue of the so called research exemption is a key one, in perception if not reality.

As the research exemption is not specified in U.S. patent law, it is open to interpretation. Here there is widespread disagreement due to the opinion in Roche Products v. Bolar Pharmaceutical Co. (Fed. Cir., 1984). However, the particular case, which as a precursor to marketing a generic drug had clear and specific commercial intentions, did not apply to experimental use so that some consider the extensive and often quoted commentary as not dicta—that is not applicable in other instances.[4] I agree with Bent (1989) in the regard that not allowing experimental use flies in the face of the evolutionary nature of inventions and the whole purpose of the patent system. Nevertheless that belief/interpretation is not widely held; many are concerned by this issue.

The concern has arisen in the proposed (but not enacted) amendment to the Patent Act (the Transgenic Animal Patent Reform Act). A 1989 workshop sponsored by the Crop Science Society of America similarly recommended a clarification of this matter, through legislation if need be (American Society of Agronomy, 1989, pp. 186-87).

There is an urgent need to clarify this matter to researchers if not in the law. Legitimate concerns exist and need to be addressed.

REVISIONS TO UPOV

UPOV is the international convention for Plant Breeders' Rights (PBR). PBR are patent-like grants with some additional exemptions (especially the right of farmers to plant seed held over from harvest) which reduces the scope of protection compared to patents. In the U.S., UPOV is implemented as the Plant Variety Protection Act of 1970.

UPOV is significant here because in Europe, and where European-type law is applied, it acts as a "separate but equal" system where UPOV covers plants and patents other inventions. This is done through the ban on so-called double protection (UPOV Article 2) from which the U.S. has an exemption.

The point of raising this is to note that major revisions have been proposed and are being discussed at this time (UPOV 1990). Like many changes, this revision promises benefits as well as possible problems for U.S. agricultural biotechnology firms. Among the major benefits, the ban on double protection is proposed to be dropped, facilitating the way for the much needed patenting of plants in Europe.[5] Of possible concern is the institution of dependency rights (Article 14 (2)–Alternative A). De-

---

[4]The case identified in Roche Products was rectified by a 1984 amendment to the U.S. Patent Act (Sec. 271(e)) giving specific rights for testing pharmaceutical products.

[5]Other approaches to this end are being taken currently–see Council Directive (OECD,1988).

pendency rights define the opportunity for the developer of the "initial" variety to authorize or not the sale, offering for sale, etc., of "essentially derived varieties." That latter term refers to changes which preserve the essential characteristics of the variety, such as variety selection, back-crossing or transformation by genetic engineering. Varieties are not to be sequentially dependent (A-B-C), but rather dependent on the initial variety (A). My concern is that the holder of the rights to "A" will gain quasi-intellectual property rights to the thousands of attributes of the variety which he/she did nothing to create. It is just this potential co-opting for individuals of plants in the public domain which is so feared and criticized by breeders and others.

At the same time, proposed rights extend only to the "products made directly from harvested material" (Article 14 (1)—Alternative A). One can imagine that the breeder of a new maize variety would be powerless to prevent the importation of foodstuffs using corn starch made from that variety. Thus the proposed revisions, at least in one alternative, appear to provide too much protection in one respect and too little in another. Yet until such time as the patenting of all plants is widely granted around the world, UPOV will remain the available form of protection for U.S. inventors in this sector. Close attention to these changes and their effects is needed immediately. This is not an abstract exercise that is underway.

CONCLUSIONS

My attempt has been to establish the state of intellectual property protection in the U.S., and especially that for agricultural biotechnology, as being quite extensive. Its shortcomings, in my view, are not principally responsible for slow commercialization of these products. That point aside, intellectual property legislation is evolving also and I have identified several areas where, in my opinion, further attention needs to be directed to achieve a better balance between private incentive and public well being.

SELECTED REFERENCES

American Society of Agronomy. 1989 . *Intellectual Property Rights Associated with Plants.* Madison, WI.

Barton, J. H. 1989. Section 337 and the International Trading System. Paper presented at the Conference on Technology, Trade and World Competition 27 Nov. 1989 Brussels and 29 Nov. London.

Bent, S. 1989. Issues and Prospects in the USA. in *Animal Patents: The Legal, Economics and Social Issues.* New York: Stockton Press, pp. 5-15.

Bent, S. A., R. L. Schwaab, D. G. Conlin and D. D. Jeffrey. 1987. *Intellectual Property Rights in Biotechnology Worldwide.* New York: Stockton Press.

Biotechnology and Plant Variety Protection Newsletter. 1988. Trade Act Finally Enacted, 3: 5-6.

Butler, L.J. and B.W. Marion. 1983. *Impacts of Patent Protection in the U.S. Seed Industry and Public Plant Breeding.* University of Wisconsin, NC+117 Monograph 16, Sept.

Dasgupta, P. and J. Stiglitz. 1980. Uncertainty, Industrial Structure and the Speed of R&D. *Bell J. Economics,* Spring.

Gershon, Diane. 1991. Biotech Patents: Amgen Scores a Knockout. *Nature* 350:99.

Herber, B. P. 1975. *Modern Public Finance. 3rd Ed.* Homewood, IL: R.D. Irwin, Inc.

Jewkes, J., D. Sauers and R. Stillerman. 1969. *The Sources of Invention.* New York: Norton. Second Ed.

Lesser, W. 1987. Grace Periods in First-to-File Countries. *European Int. Property Rev.,* 3(Nov.): 81-85.

Nogués, J. 1990. *Patents and Pharmaceutical Drugs: Understanding the Pressures on Developing Countries.* Washington, DC: The World Bank, WPS 502, Sept.

OECD. 1988. Proposal for a Council Directive on the Legal Protection of Biotechnological Inventions. COM (88) 496 final—SYN 159, Brussels, 17 October.

Ratner, Mark. 1991. Patent Litigation: EPO, Factor VIII Shockers *Biotechnology* 9: 327

Scherer, F.M. 1980. *Industrial Market Structure and Economic Performance. 2nd Ed.* Chicago: Rand McNally.

Siebeck, W.E. ed. 1990. *Strengthening Protection of Intellectual Property in Developing Countries: A Survey of the Literature.* Washington, DC: The World Bank, Discussion Papers 112.

UPOV. 1990. Basic Proposal for a New Act of the International Convention for the Protection of New Variety of Plants. DC/91/3, Geneva, Nov. 9.

U.S. Congress, Transgenic Animal Patent Reform Act. (HR 4970).

Van Horn, C. 1987. Recent Developments in the Patenting of Biotechnology in the United States, In proceedings of the *Symposium on the Protection of Biotechnological Inventions*, Ithaca, NY, June 4-5, pp. 73-85.

COMMERCIALIZING AGRICULTURAL BIOTECHNOLOGY

ROGER SALQUIST

Chair and CEO
Calgene Inc.
1920 Fifth Street
Davis CA 95616

Calgene, an international agribusiness biotechnology company, is developing new varieties of edible oil crops, alfalfa and tomato seed and seed potato tubers.

Some time ago I decided that every time I hear Jerry Caulder speak, I would remember and try to use one of his quotations. I think it appropriate to repeat his quote from Will Rogers: "It's not what we don't know that hurts us, it's what we know that ain't true that's gonna kill us in the end"(sic). As we talk about the real risks and rewards in biotechnology, it behooves us to remember those words.

I will address the impediments to industrialization and commericalization of agricultural biotechnology. One of the goals of the meeting was to assess "the reasons why many biotechnology innovations have failed to develop as predicted." Well, I would dispute that view. Biotechnology innovations have come almost exactly in line with what responsible and knowledgeable people involved in this field since the late 1970s and early 1980s have predicted. For instance in the early 1980s, Martin Apple, president of International Plant Research Institute, was widely quoted—whether accurately or inaccurately—predicting plant biotechnology would generate pork chops on trees. If that is the standard as to our progress, obviously we are not there. On the other hand, Tom Urban, president of Pioneer-Hi Bred, the largest seed company in the world and one of the more knowledgeable individuals in this field, still predicts we will not have genetically engineered plant products on the market and making an impact until the year 2000. In many respects, Tom Urban is as wrong as Martin Apple was as to what is going to happen. We are on the verge of having a whole plethora of products that are going to have material, economic and positive environmental impacts on the agricultural arena.

Let me comment on a few of the specific barriers to commercialization and the standing of a few specific products in the regulatory process. Earlier in this volume (p. 111), Bob Nicholas presented the regulatory background. I will comment on where we stand in the process, using examples from Calgene, my own company, because those are the most familiar ones. To date, the U. S. Department of Agriculture (USDA) has issued over 120 permits to conduct field trials with genetically engineered plants. Calgene was the first company to get such an approval in November, 1987, and has since received over 20 such permits. The average time from the day we file with the USDA until we get those approvals is 105 days. That is very reasonable in the context of any federal bureaucracy, particularly the regulatory ones. Calgene's most recent field trial represents work we are doing in genetically engineered cotton which is in the third year of field trials. Under a single permit, we received approval to conduct 34 trials in 12 different states. All of those trials were planted as of the last Friday in May, despite the highest amount of rain in the southern U.S. since Noah built the ark. Those trials will generate data that will not only answer the questions of the safety of the plants themselves, but that data will be shared with responsible researchers in various university systems and within the Food and Drug Administration (FDA) itself. That data will allow the USDA and other regulatory agencies to make appropriate determinations of risks, if any, in going to full commercialization of those types of plants.

We are on the verge of having a whole plethora of products that are going to have material, economic and positive environmental impacts on the agricultural arena.

The FDA has taken a lot of criticism, and I feel those criticisms are completely unfounded and inappropriate. In fact, FDA has moved forward with this type of technology assessment. Before FDA can decide what regulations they want to issue, they need to review specific data, understand what is involved and what changes really occur in plants.They have to review information that has been generated,. And that is happening. Calgene filed a petition for the use of a selectable marker and vector in November 1990, asking FDA to review the safety of 80 percent of the plasmid rather than just a specific target gene, in order to separate the issues. There are lots of different ways to file data with FDA, but Calgene specifically selected the advisory opinion route because it is a public process that invites public comment, and because every single piece of data Calgene submitted to FDA is public. Anyone that wants it can have it, either from Calgene or FDA.

In fact on May 1, 1991, in the *Federal Register*, the FDA published a request for public comment. Calgene entered the 90-day period for comments on the safety of the use of the technology. I encourage all of you that have an abiding interest in this to comment. Comments are exactly what FDA wants, and part of this whole process is getting public input. I hope all of you will avail yourselves of that opportunity—the comment period

FDA has the technical competency... to make the scientific assessments necessary to assure the American people that this technology is not only safe but is beneficial.

closes on July 30. My strong understanding and belief is that FDA will, on the basis of the focused thinking they have done on this topic now, issue some specific guidelines or points to consider. I do not know what the exact format will be, but I expect a decision from FDA before the end of 1991. I believe you will see FDA making a positive determination on the safety of genetically engineered whole foods in the calendar year 1992–a forecast I am totally prepared to stick with.

Next let me comment on agency collaboration–one of the issues in the public arena where people like to stir up trouble. There is the impression that massive gang warfare is going on between various agencies back in Washington, which is not true (with one notable exception that I will avoid). Certainly, in my mind, the collaboration between FDA and USDA is excellent. For the last five years, numerous meetings have convinced me that FDA has the technical competency through all of their various Centers of Food Safety, Center for Veterinary Medicine and various departments of toxicologists and physiologists, to make the scientific assessments necessary to assure the American people that this technology is not only safe but is beneficial.

There is not a single plant biotechnology product to date that has been delayed by the regulatory process.

Let us focus on the major structural impediments to agricultural biotechnology. First and foremost is the nature of the science. There is not a single plant biotechnology product to date that has been delayed by the regulatory process. Plant products have been delayed by the nature of the science involved, the inherent difficulty and the time required to work with plants. I would not make the same statement with microbes, but I assert that with plants, you are talking about both recalcitrant and slow moving science. For example, if you are going to genetically engineer a tomato, the shortest period of time that you can take is about six weeks. And that just gets you to the plantlet. You must still grow the plant, get the seeds and harvest the seeds to determine whether you have affected the reproductive capacity of the plant. Thus, the science is the principal problem. It takes a long time. You can only do one experiment a year. You transform a plant, grow it up and get the results a year later, which tends to make progress move rather slowly. The inherent nature of the science and the fact that you are dealing with the plants is an impediment. A second impediment is the whole base of knowledge in this field. Previous speakers have talked about it and the small allocations from the federal research budget is yet another impediment. Federal research spending in the plant sciences or agricultural sciences is miniscule. It does not even make the chart compared to the money that has been historically, and is currently, piled into biomedical research. Jerry Caulder estimated that two percent of the total federal research budget goes into agricultural science as broadly defined. Another measure to consider: historically, the National Institute of Health (NIH) has poured approximately three billion dollars a year into research in the human area. In contrast, over three years ago my good friend Paul Stumpf, who heads the competitive grants program at the USDA, was talking about a $15 million budget in terms of competitive grants. Now it is $75 million. That is still a drop in the bucket compared to what is being spent in the development of basic knowledge in other areas. Just to give you an idea of the parameters, I calculated that private industry in the U.S. invests $350 million a year on plant biotechnology research. That is a substantial figure, although again it is not relative to the base of knowledge that we have and need to know. Calgene spends about $12 million a year and has invested $75 million in research in this area in the last ten years. The nature of the science and level of knowledge are major barriers to accelerated progress.

The nature of the science and level of knowledge are barriers to accelerated progress.

A third major barrier to commercialization and to making an impact with this technology is finance. It is damn tough to raise money in any agricultural enterprise in the U.S. for several reasons, not the least of which is that there is no major business school that serves as a breeding ground for all the investment bankers and the financial analysts in the U.S. Virtually no business schools has courses with agribusiness in the title, let alone agribusiness in the curriculum. So the financial decision makers in this country are totally ignorant of the underlying economics and opportunities presented by the largest single industry in the country. That is why when you go to a Wall Street banker and say, "I've got a great ag/biotech idea and need money", they just sort of stare out the window and look at the clouds because they do not understand anything about it. It is too difficult for them to take time to learn—they would much rather find a guy who has got a cure for AIDS. Then they would say, "Great, I'll finance it. Don't give me any details. It sounds great. Go for it." There is a huge knowledge gap out there.

The financial decision makers in this country are totally ignorant of the underlying economics and opportunities presented by the largest single industry in the country.

Some finance problems are related to frustrations stemming from a failure to meet expectations. Agricultural biotechnology is highly visible.

...there will be no more agricultural biotechnology start ups in this country in the foreseeable future...

Typically, the kinds of products we are talking about here and that other companies in the industry are talking about have been developed in the bowels of large corporate research groups where they have had a ten year gestation period before seeing the public light. Now what has happened is that the only way companies such as Calgene and other agricultural biotechnology companies could raise money to do research was to go public with the prospects of what they were doing at a very, very early stage. People saw for the very first time how truly difficult and long–term science is. Historically, the development of a drug has never become visible to the public until it is in a phase three clinical or has actually been approved. What the public does not see is the ten years required to move a drug along and get it to clinical trials. Public visibility is a cross that we have to bear, a fact of life which increases the difficulty of financing.

I will make a prediction—that there will be no more agricultural biotechnology start ups in this country in the foreseeable future, because it will be impossible to finance them. There are only about half a dozen successful agricultural biotechnology companies at the present time, and I predict at least half of those will be out of business or be acquired by foreign companies within the next two years. That is a particularly chilling message. While while lots of people in the U.S. are running around worrying about the demise of the small family farm (which is a socioeconomic phenomenon that is going to happen irrespective of technology) they fail to realize that nine of the top ten or twelve companies, representing over 90 percent of the agrichemical sales in this country, are owned by foreign corporations. Three of the top five seed companies are owned by foreign corporations. If you are worried about a narrow–minded, inward–looking and self–perpetuating focus of technology, there is no better way to have it happen than to have all of your inputs controlled by mega corporations that do not have their roots in this country. People need to start learning a little bit about financing and fostering innovation—instead of stifling it. Otherwise we face some very chilling prospects.

The next issue is the whole convoluted structural situation in international agriculture–the common agricultural policy in Europe, the Farm Bill here. The structural situation here unequivocally discriminates against innovation. New practices are discriminated against, because of the existing political and financial structure. Canola is a good example. Canola is a new crop in the U.S. there are 40 million acres are grown elsewhere in the world, but it is new to the U.S. It is easy to show that canola grown in the winter is the only crop which you can introduce as a diversion crop for crop rotation. It is the only alternative to winter wheat in huge portions of the U.S. Even with the new, non–proven varieties we have, you can show unequivocally that a canola farmer can generate $30 to $50 an acre more profit growing canola than he can growing winter wheat. But the Farm Bill gives farmers a $40 per acre subsidy and takes away any incentive for the farmer to innovate. Farmers say "I don't care, why should I innovate?" This is a very serious structural problem. We are making a little bit of headway; the 1990 Farm Bill provides a little more flexibility and that is helpful. But we still have a long way to go.

We have had 15 years of successful research and of research expansion without a single person in the U.S. getting ill from this technology. That is a safety record unmatched by any other industry in the history of the world.

The final impediment is public acceptance. I echo Walter Truett Anderson's final point — we must learn from experience. Rather than reflecting on the Natuffians' experience as Walter did, I thought about our experience with fruit flies (*Drosophila*). Last week, in the *Wall Street Journal* there was a front page article on *Drosophila* research being undertaken by William Quinn at Massachusetts Institute of Technology. Quinn is particularly interested in fruit flies because the molecular and genetic structure of the *Drosophila* brain is very similar to the human brain. There are lots of parallels and it has been discovered that the fruit fly has developed a learned response. The lowly fruit fly can learn from experience. However, Dr. Quinn is studying a particular mutant group who cannot do so. I would place many of the adamant opponents of agricultural biotechnology in the category of the abnormal fly that Dr. Quinn is looking at, because they simply seem unable to learn from experience. We have had 15 years of successful research and of research expansion without a single person in the U.S. getting ill from this technology. That is a safety record unmatched by any other industry in the history of the world. Even in the wheelbarrow industry, I am sure a falling wheelbarrow killed someone. People have to put this safety record into perspective. The reason we are doing field trials is to learn. We are generating data that allows us to move forward without worrying that the sky is going to fall on us. I think we all should keep that in mind and move forward aggressively to truly reap the harvest of this great technology.

THE PUBLIC SECTOR'S ROLE IN BIOTECHNOLOGY RESEARCH

The concept that the public sector has a major role to play in research — and particularly in agricultural research — is an historic one. It was part of the original mandate for the Department of Agriculture (USDA) in 1862, but perhaps its strongest expression came in 1887 with the passage of the Hatch Act to establish the land-grant system of universities. Recognizing that it is in the public interest to supply funds to finance research, the land-grant philosophy of research, education, and public service has been the foundation of America's agricultural productivity for nearly 125 years.

CHARLES E. HESS

Assistant Secretary for Science and Education
USDA Room 217–W
Administration Bldg.
Washington, DC 20250

Hess is responsible for research and education programs in the food and agricultural sciences, including planning, evaluation and coordination of state and federal activities.

Thus the acknowledgment by government — and the public — that agricultural research is a valuable use of tax dollars has an historic basis — and a foundation that has continued to be solid over the years as the investment is paid back many times in vital contributions to agriculture, to the consumer, and to the national economy.

Overall, in the United States, the government has tended to take on the role of funding more fundamental research. Thus, not every project will have a clear end product, but may be geared toward increasing our knowledge about the functioning of organisms at the cellular and molecular level. As a result, the public sector has played, and will continue to play, a major role in doing fundamental research which contributes to the foundation of the biotechnology industry.

High-risk research is also an area which is ripe for public funding. As part of their development, biotechnology firms are going through an evolution. Following an initial large investment of venture capital, there were hopes for the quick generation of substantial returns. But as the first flush of excitement and expectation has died down, reality has set in in terms of the time needed to get products out into the marketplace.

As companies become more market-oriented, universities and government are playing an increasingly central role in doing high-risk work. For example, not too many biotechnology firms are doing salt tolerance or drought tolerance research because there are uncertainties about whether it will work. Therefore, research on this important trait has fallen to the public sector.

In addition, I do not want to overlook the equally significant responsibility of the universities in training the young men and women who will become employees in the biotechnology industry and faculty in the universities. To meet the growing demand for qualified people to guide future technologies, we must attract top-notch young men and women to science and agriculture, train them well, and endeavor to increase the racial, cultural, and gender diversity of that brain pool. Young people are the lifeblood of any long-term research endeavor.

To meet the growing demand for qualified people to guide future technologies, we must attract top-notch young men and women to science and agriculture, train them well, and endeavor to increase the racial, cultural, and gender diversity of that brain pool.

TECHNOLOGY TRANSFER

There is growing concern that although the U.S. is preeminent in basic research we fall behind other countries in turning that knowledge into products. While the movement of graduate and postdoctoral students from the university to industry has always been one way in which technology transfer is accomplished, the passage of the Federal Technology Transfer Act in 1986 was a major advance. It clearly established the appropriateness of government scientists working closely with industry to facilitate application of their research results.

Not only is the Agricultural Research Service (ARS) aggressively implementing the Act's provisions, but USDA has been directed by Congress to increase developmental research within ARS. In addition, I chair a task force which is developing specific mechanisms for promoting closer cooperative interaction among federal, state, university, and industrial scientists. Our goal is to facilitate the utilization of discoveries and inventions originating in USDA and State Agricultural Experiment Station Laboratories for the benefit of society.

REGULATORY UNCERTAINTY

The public sector has a role in two major issues that will determine the future of biotechnology: 1) regulatory uncertainty and 2) public perception.

We have been working for the past year on developing a definition of the scope of organisms which should receive regulatory oversight. The publication of guidelines for field testing of "organisms with deliberately modified hereditary traits" — genetically modified organisms — was an example of how this definition could be used.

The regulatory framework for biotechnology should be based on sound scientific principles, in which oversight is commensurate with the level of risk. We have three major goals: 1) to avoid singling out recombinant-DNA technology as being any more risky than other procedures used to modify an organism; 2) to refrain from unduly hindering research with burdensome and unnecessary overregulation; and 3) to provide assurance to the public that there is adequate review prior to the release of modified organisms if we are unfamiliar with how they will affect the environment or health.

PUBLIC PERCEPTION

Some people feel that we should not use the term "genetically modified organisms" because it indicates there is risk. The issue is that there is already a public perception of risk — and to much of the public, perception is reality. This issue must be resolved or else the great potential of biotechnology will not be realized.

In terms of the public perception, it is part of our role as scientists and end users of technology to get across to the public the facts it needs to make informed decisions — to help people look at the big picture and weigh the long-term benefits and costs. I take exception to those who would paint biotechnology as competing with sustainable agriculture for funding. Sustainable agriculture is a systems approach to producing food and fiber efficiently, economically, and in harmony with the environment. As a systems approach, it needs input from ecologists, soil scientists, computer scientists, systems engineers, and economists.

In terms of the public perception, it is part of our role as scientists and end users of technology to get across to the public the facts it needs to make informed decisions — to help people look at the big picture and weigh the long-term benefits and costs.

It also needs molecular biologists and biotechnologists who now have the tools to ask questions about the molecular and biochemical basis of desirable traits such as disease and insect resistance and photosynthetic efficiency (we currently use less than one percent of the sun's radiation). In addition, we can delete components in food that are detrimental to health and add others that are beneficial. Thus, biotechnology can help agriculture be sustainable, productive and nutritious.

Another mechanism for addressing public concerns is risk assessment research. For the first time, the Farm Bill explicitly directs USDA to support biotechnology risk assessment research. The bill directs the Secretary of Agriculture to establish a grant program to fund research on methods to confine introduced organisms, monitor their dispersal, study potential gene transfer, and investigate other areas in which biosafety information may be incomplete.

To support this research, USDA will designate one percent of its biotechnology research funding exclusively for risk assessment work. Although the Department is still looking at budget data and discussing the details of implementing this legislation, it is likely that the funding level for risk assessment research will be about $1 million a year.

FUNDING

In closing, I would like to address the issue of funding for biotechnology research. In 1987, the National Research Council (NRC) published a report on *Agricultural Biotechnology — Strategies for National Competitiveness* which recommended that in order for agricultural biotechnology to reach its full potential, a $500 million investment should be made in fundamental research in the agricultural biological sciences.

Congressman George Brown picked up on this recommendation and suggested a national institute of agriculture funded at the $500 million level. Then, in 1989, the Board on Agriculture of the National Research Council published a report entitled *Investing in Research — A Proposal to Strengthen the Agricultural Food and Environmental System.*

These two NRC reports formed the basis for the National Research Initiative on Agriculture, Food, and the Environment. This initiative was launched in the President's FY 1991 budget with a recommendation for funding in its first year at $100 million and a commitment to add $50 million in each of the outyears — provided that funds were appropriated on a non-earmarked basis. In the same year, Congress authorized funding of the initiative at the recommended level of $500 million per year, reaching that amount of funding by 1995.

As you know, the authorization and appropriations process are two distinct activities. In FY 1991, Congress appropriated $73 million, and in the

FY 1992 budget currently being considered by the appropriations committees, the President has recommended funding at $125 million. In addition, there is a $25 million competitive facilities program which is an attempt to address the earmarking problem.

Although not all of these funds are dedicated to biotechnology, a considerable amount in the plant and animal systems areas will support basic work needed to effectively use the tools of biotechnology. In the $35 million for plant systems, $11 million is dedicated to map genes that regulate agriculturally important traits such as insect and disease resistance and drought tolerance. In the $53 million recommended in FY 1992, $18 million is recommended for genetic mapping.

FCCSET

There is also a new federal approach to research funding through the Office of Science and Technology Policy (OSTP) which operates the Federal Coordinating Committee of Science and Engineering Technology (FCCSET). The purpose of FCCSET is to bring together agencies across the government which have interest and expertise in a particular priority area — for example, global climate change. By "comparing notes," so to speak, we establish a baseline picture of what is going on. We can see the overlaps — and the gaps.

Just as a doctor would use an X-ray to guide her in choosing a treatment, the baseline picture helps us develop a research plan. Each agency then develops that part of the plan in which it can make the best contribution. Furthermore, by bringing the Office of Management and Budget (OMB) into the planning stage, we can assure that each group's part of the overall plan does become part of that agency's budget. FCCSET is a successful strategy to coordinate research planning and budgeting and to establish national research priorities.

Last year, in addition to global climate change, FCCSET added two more areas to the planning process: high-performance computing, and science and mathematics education. While across government there is an overall attempt to keep the budget less than inflation, all three initiatives received increased funding in the President's FY 1992 budget. When FCCSET met a few weeks ago, two other candidates were proposed for the planning process: advanced materials and processes, and biotechnology.

We will start out doing a "crosscut" to determine the current level of funding and then plan for the outyears of 1993 to 1995. It is too early to know whether this exercise will turn into an initiative as did global climate change, high-performance computing, and science and mathematics education, but this is the first step — which may then lead to a component in the President's 1993 budget.

A strong national commitment to public funding of agricultural research, technology, and education is essential to the short- and long-term interests of both the U.S. and the world.

CONCLUSION

From ancient civilizations to our current technologically advanced society, national leaders have understood that new scientific knowledge can be a tremendous instrument of national strength and public good.

A strong national commitment to public funding of agricultural research, technology, and education is essential to the short- and long-term interests of both the U.S. and the world. Secretary of Agriculture Ed Madigan has stated (House testimony 4/17/91) that "like other sectors of the economy, agriculture is increasingly dependent upon technological advances to ... meet immediate and long-term challenges...."

The role of the public sector in helping agricultural research meet those challenges has historically been a central one — and it will surely remain so. I foresee the strong continuation of investment in publicly-funded agricultural biotechnology research in the United States, which I feel is essential to achieve the goals of an economically and environmentally sustainable agriculture.

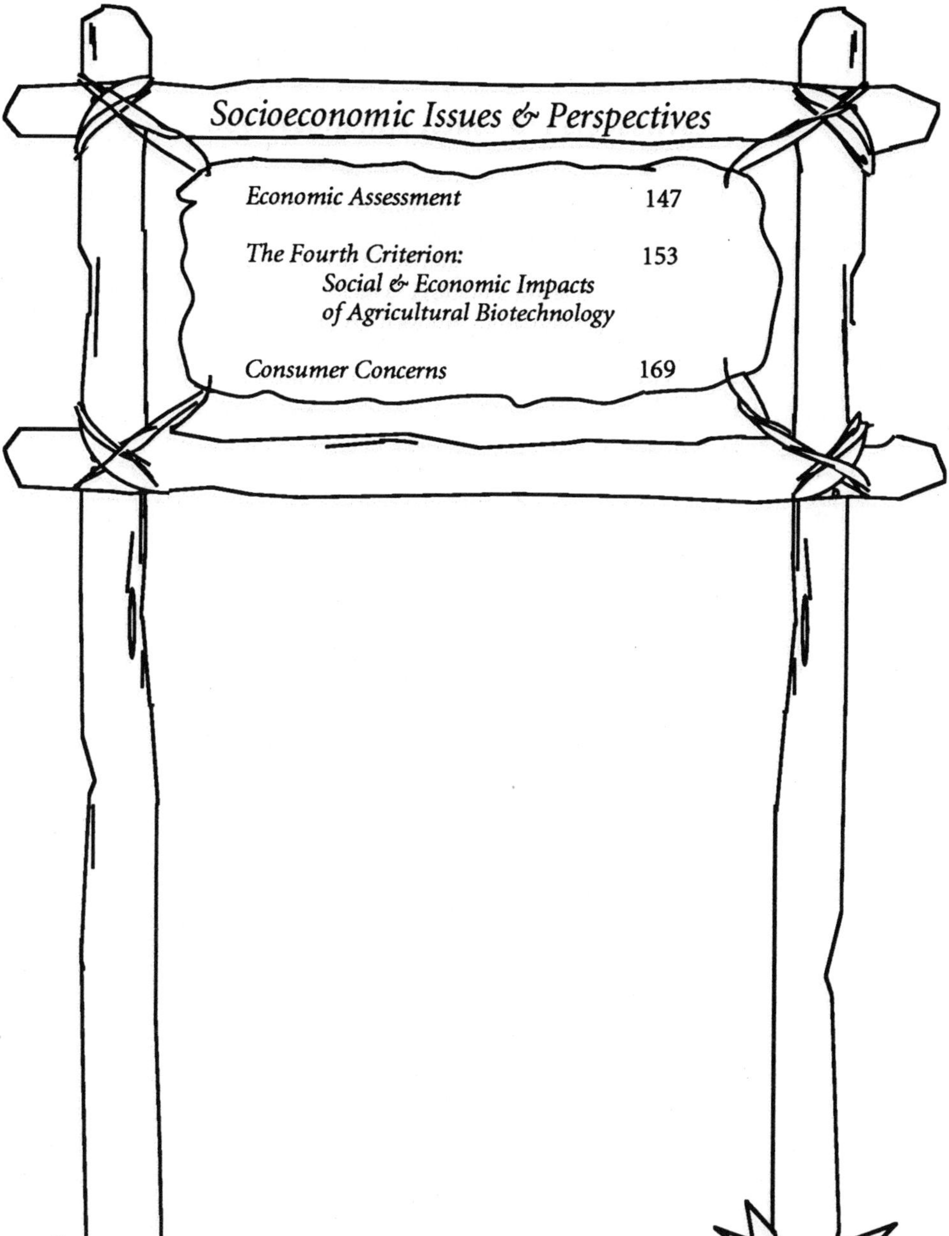
Socioeconomic Issues & Perspectives
Economic Assessment 147
The Fourth Criterion: Social & Economic Impacts of Agricultural Biotechnology 153
Consumer Concerns 169

ECONOMIC ASSESSMENT OF AGRICULTURAL BIOTECHNOLOGY

In predicting the impacts of biotechnology on agriculture, the future is seen through a glass so dark that only the most striking features of the economic landscape are visible, and these features themselves are changing. Any economic assessment of agricultural biotechnology, then, must take account of these changes, whose nature can now be fairly well discerned.

SUSAN OFFUTT

Office of Management and Budget
Executive Office of the President
Washington, DC 20503

Formerly a faculty member at the University of Illinois and an agricultural economist at the Department of Agriculture, Offutt has been Chief of the Agriculture Branch of the National Resources Division at OMB since 1988.

To date, much of the analysis of biotechnology has focused on what is called the "microeconomics" of farm and food markets. The focus has been on the individual unit of production, seeking to determine the response of the atomistic farmer or firm presented with the opportunity (perhaps the imperative) to adopt a new technology. Neoclassical economic theory allows analysts to make some fairly robust predictions about the adoption decisions of these individual units, *given no other changes in the economic environment.*

As the decade of the 1980s passed, however, the assumption of constancy in what can be considered the "macroeconomics" of agricultural markets has become less tenable. In fact, today the outlines of major change in public policy can be seen that will profoundly affect the widespread adoption and effects of agricultural biotechnology. The anticipated successful conclusion of the multilateral trade talks and the coming reworking of the laws governing pollution of the nation's waters will combine to work distinct changes on the incentives and constraints on the farm economy.

Evolving attitudes about nutrition and health as well as life-styles will further affect the use of biotechnology, particularly in food processing. In the years ahead, the most welcome contribution biotechnology could make would be in allaying (not stimulating) consumers' concerns about food safety.

In the years ahead, the most welcome contribution biotechnology could make would be in allaying (not stimulating) consumers' concerns about food safety.

The present analysis puts this important aspect of the economic assessment of biotechnology aside, however, and concentrates instead on an assessment of the environment faced by production agriculture.

SOURCES OF MACROECONOMIC CHANGE

A new international farm trade regime and domestic environmental policy focused on farming will profoundly change the environment in which new products and processes are adopted. In this respect, biotechnology will not be the tail that wags the dog, although an exclusive focus on the microeconomics of the adoption process may have contributed to such a perception. For a farmer, the choice of biotechnologies is but one in a set of input decisions that are conditioned by the prices of his output and other inputs. The trade and environmental changes in the 1990s will affect both market prices and the implicit price of use of the natural environment.

Farm Support Policy — The recent Congressional affirmation of the President's authority and flexibility to negotiate makes quite probable a successful conclusion to the Uruguay round of trade talks. Under the auspices of the General Agreement on Tariffs and Trade (GATT), the U.S. will be pushing a new farm policy regime that does not distort international commodity trade. While direct income subsidies would likely be permitted, the current configuration of supply control and price support for specific commodities would be phased out. At the same time, domestic U.S. budget realities may well require further reductions in the level of support, even though the level of non-distorting subsidy is not subject to GATT discipline.

Taken together, the move away from commodity-specific subsidies based on the level of production and a reduction in the overall level of support returns a large measure of the responsibility for managing the risks of farming to the farmer. Over the past 60 years, the Federal government has assumed an ever-greater share of this risk through, for example, the imposition of supply control and price support, the subsidized provision of crop insurance and disaster assistance, and the stockpiling of surplus crops. By these means, government intervention has both moderated and absorbed much of the instability inherent in agricultural production.

The provisions of the recently-enacted farm bill provide a clue to the future. The reduction in the level of support was imposed by scaling back the number of acres on which production is eligible for price support. A farmer now has the "flexibility" to plant for market on the acres not receiving subsidies. In many cases, the planting choice will be a crop other than that the subsidy program supports. A move to an income-based subsidy would accelerate this trend away from support based on production of a specific commodity.

Diversification across crops and perhaps livestock enterprises will be a farmer's most logical response to change and reduction in the domestic policy regime. Federal commodity programs have historically encouraged a farmer to put all his eggs in one basket, but future subsidies will not so favor the production of one set of "basic" commodities over others. By moving away from specialization and toward reliance on multiple outputs, the farmer will hedge his bets with nature and the commodity markets. His input choices will be conditioned not only by prices but by the input's effect on the stability of his outputs. The farmer is now managing his farm enterprise as he would a portfolio of stocks, considering both risk and return.

By moving away from specialization and toward reliance on multiple outputs, the farmer will hedge his bets with nature and the commodity markets.

Environmental Policy —The Clean Water Act is due for reauthorization during this session of the Congress. The revision of the law protecting the nation's waters will focus on the contribution of farming to surface water quality degradation. Over the years, other "point" sources of water pollution, such as factory waste pipes, have been brought under control. Agriculture, as the major "nonpoint" source, has gone unregulated largely because of the difficulty of identifying individual sources of pollution. However, the Congress is now apparently feeling up to the challenge.

At present, the means for controlling farm pollution of water are under scrutiny. The Administration has supported a course of pollution prevention. The President's Water Quality Initiative has endorsed the subsidy of the development of environmentally "benign" farming techniques and promoted their transfer to farmers. And, the Administration's proposal for the 1990 farm bill gave farmers total flexibility in planting on program acres, thereby promoting crop rotation as a straightforward means of minimizing jeopardy to the environment posed by monoculture. Farmers tend to prefer an approach under which they receive incentive payments to adopt certain practices, but budget constraints make this an unlikely outcome. Another alternative would mandate (through regulation) prac-

All input and output choices are likely to be affected when the price of using the environment as an input increases.

tices expected to be less harmful to water quality. Trading of the right to pollute, as with provisions of the Clean Air Act, might be another avenue to pollution control.

Even without being able to describe the outcome of the Clean Water debate in detail, a prediction of the impact on farmers can be made. The price of the using the natural environment in farm production will increase. One way or another the pollution of water will become more expensive to farmers, who may find their production costs raised by the prohibition against the use of certain techniques, nutrients, and/or chemicals. Historically, the environment has been valued at zero in farm production; however it does so, the Clean Water Act will change that.

All input and output choices are likely to be affected when the price of using the environment as an input increases. A recent article in the *New York Times* business section underscored this point in promoting the stocks of biotechnology companies working on biopesticides. Preventing water pollution by either nutrients or chemical pesticides may also push farming away from specialization, just as changes in the subsidy regime can be expected to do. Then, new crop inputs may have to function in new rotations, and new animal inputs may no longer be used in today's large confinement systems but in integrated crop/livestock production systems.

AGRICULTURAL BIOTECHNOLOGIES IN A NEW CONTEXT

In the next decade, anticipated changes in domestic farm support and environmental policies will have profound effect on the macroeconomics of agriculture. Two major implications for the use of biotechnologies emerge: first, the need to anticipate the demand for inputs that help farmers manage risk and second, the need to prevent environmental pollution.

Current signs point to the re-emergence of diversified farming as the hallmark of American agriculture. While a return to the small scale farms of the 1930s is unlikely, more diversified and regionally concentrated farm production may well be in the future. As examples, for both economic and environmental reasons, extensive cropping of the Northern Plains and concentrated livestock production throughout the country may become impractical. In looking to the future, analysts should consider predictions about the outcome of the GATT negotiations as a starting point. And, by considering the tenants of sustainable agriculture, the virtues of integrated systems in a new context may become clearer.

To this point, economic assessment of biotechnology has concentrated on the current context of farm production. But, as time passes and biotechnologies begin to come on line, the policies that condition farmers' responses will become very different from those in evidence today. In particular, the farmer's need to manage more of the risk of farming than ever before adds another dimension to his decision-making. The contribution of input choice to output stability becomes at least as important as the effect on the level of output. The theoretical framework for analysis of production choices under risk is fairly well-developed; it now needs to be applied to the consideration of individual biotechnologies.

...as time passes and biotechnologies begin to come on line, the policies that condition farmers' responses will become very different from those in evidence today.

At the same time, modifications to environmental laws will make farming more expensive, most likely through restriction of input choices and techniques. Already, there is a general awareness of the contribution biotechnologies can make to solving pollution problems. Public support of research for biotechnology might well find its biggest payoff by making environmental sensitivity a high priority. To the extent that plants' needs for added nutrients and pest protection can be moderated, the environment could be better protected.

CONCLUSIONS

Facing the future with biotechnology requires recognition that the times are changing and agriculture with it. To the extent that American agriculture will look and act differently, its willingness and ability to adopt biotechnology may be affected. Economic assessment must consider the effects of changes in the overall "macro" environment in making predictions based on the "micro" unit. These remarks have offered a rough guide to the formulation of that analysis of the future.

Would acknowledgment of the changes in store affect the development of biotechnologies underway today? It is hard to say, but one can say with some confidence that that knowledge will certainly determine the successful marketing of products tomorrow. The prediction of future change is put forth together with the hypothesis that it does matter to the form and function of future agricultural biotechnologies. How it may matter and to what extent, is left to those with more extensive understanding and experience.

No look toward the future would be complete, however, if it focused exclusively on the economic variables because they themselves reflect society's values and concerns. To that end, if biotechnology can be shown to make a positive contribution to the process of change in agriculture, its future will be assured. But to the extent that adoption of biotechnology is promoted as an end in itself (as an application of frontier technology) or simply as a problematic successor to conventional technology, society may raise barriers to its adoption regardless of it is other merits. Looking outward, the largely scientific community that understands biotechnology must describe its potential to help in terms the rest of society understands.

THE FOURTH CRITERION
SOCIAL AND ECONOMIC IMPACTS OF AGRICULTURAL BIOTECHNOLOGY

The world economy is currently undergoing major structural changes. A central factor in these changes has been the development and diffusion of fundamentally new technologies, in particular computers and the "new biotechnologies." Social and economic changes that result from these profoundly enhanced capacities in science and technology are visible in every sphere of human life from health, transportation and communication, to agriculture and the food supply. However, each change is associated not only with new benefits but also with new risks, latent complications and long term consequences which are often poorly understood.

WILLIAM B. LACY

Asst. Dean for Research
College of Agriculture
The Pennsylvania State University
University Park, PA 16802

Lacy, the Assistant Dean for Research and Professor of Rural Sociology, has co-authored or co-edited over fifty articles and chapters on agricultural research.

LAWRENCE BUSCH

Sociology
Michigan State University
E. Lansing, MI 48824-1111

Busch, a Professor of Sociology, has co-authored or co-edited six books on agricultural research policy and currently serves on the scientific advisory board of CIRAD, the French agricultural scientific cooperation agency.

Commercial applications of biotechnology to agriculture are still in their infancy. Despite enormous optimism in the scientific community, in national and state governments, and in the private sector, most of the products, processes, and impacts of biotechnology, particularly in agriculture, remain promises for the future. The applications of biotechnology are fraught with concern and controversy within both the scientific community and the broader public.

Often public concerns have centered on health and environmental safety issues. Similarly in Europe, the three standard criteria, human safety, animal safety, and efficacy, have been utilized to evaluate and approve new products and processes. Currently, a fourth criterion, the social and economic effects of the product or technology is being proposed. This criterion has been employed for such actions as the European Common Market's ban on growth hormones in

It may be more appropriate in evaluating research agendas and new products and processes to consider the broad socioeconomic effects as the first criterion.

food products. Recently, the Advocate General of the Court of Justice in the European Communities released his opinion on the legality of the hormone ban stating that "It was appropriate and justifiable to prohibit the administration of the five substances for fattening purposes, even in the absence of scientific evidence showing that they were harmful. A total prohibition was the only solution which could bring an end to the distortions of competition and barriers to intra-Community trade in meat, eliminate risk to public health, even if they were purely hypothetical ones, and avoid a further reduction in consumption." (Weber, 1990: 1)

Similar efforts are underway to utilize this approach as a precedent to alter the product approval process in the United States. Scientists, regulators, industry representatives, and the public in general have all voiced concern that all technologies, including genetically engineered organisms and products, could have adverse impacts. As a result of these concerns and experiences with previous technologies, an increasingly accepted position among technology assessment professionals is that: 1 - all technologies have multiple effects; 2 - many of these effects are potentially harmful and thus require conscious decisions; and 3 - these critical decisions entail moral as well scientific analysis.

Although introduced as the fourth criterion, it may be more appropriate in evaluating research agendas and new products and processes to consider the broad socioeconomic effects as the first criterion. As most scientists and policy analysts acknowledge, biotechnologies are the tools and means to achieve particular socioeconomic goals. As such they should be framed and evaluated in terms of those social goals and values.

In a democracy, the public has an obligation and a right to shape the developments of technology in terms of the broad social and economic values of their respective society. In the case of biotechnology, as we have seen, the public is increasingly exercising that obligation and right. Therefore, it is important for scientists, regulators, and policy-makers to understand and evaluate not only elements of human safety, animal safety, environmental risk, and efficacy, but also the range of socioeconomic impacts and concerns.

The potential social and economic impacts of biotechnology on the food and fiber system and society are just emerging. Consequently, the proposed implications of biotechnology for the system represent only possible scenarios. The socioeconomic effects may include impacts on: 1 - science, 2 - farmers and rural communities, 3 - consumers, 4 - the structure of agribusiness and industry, and 5 - the global market and developing countries.

IMPACT ON SCIENCE

Perhaps the most dramatic immediate impact of the new biotechnologies has been on science itself. While some argue that biotechnology is a continuation of the application of biological techniques to improve plants, animal, and microorganisms, molecular biologists contend that "biotechnology has revolutionized biology and is destined to have even greater impact than the industrial revolution on agriculture and the food system in this country" (Harlander, *et al.*, 1991).

The knowledge and tools generated by molecular biology and biotechnology have stimulated a great deal of enthusiasm and redirected large sums of money in an effort to pursue knowledge in this area. At the federal level, financial support for biotechnology has grown steadily since the mid 1980s, reaching $3.8 billion in 1991. The President's budget for 1992 calls for an 11 percent increase to $4.1 billion. While 80 percent of the federal budget has been devoted to the National Institutes of Health program, support for agricultural biotechnology has been relatively meager, constituting less than three percent of federal expenditures for biotechnology. There is, however, significant optimism that agricultural biotechnology will receive substantial increases at the federal level through the National Research Initiatives Program. In addition, industrial expenditures for biotechnology research and development had grown to $2 billion by 1990, with a large portion of the expenditures devoted to agricultural biotechnology (Metheny and Monahan, 1991).

The techniques and tools of biotechnology are facilitating basic research efforts to understand the intricate, complex, functioning of living organisms at their molecular and cellular level. Molecular biology in conjunction with other basic research is accelerating the accumulation of knowledge in traditional disciplines such as biology, genetics, plant physiology, and biochemistry. Moreover, biotechnology, particularly in agriculture, may truncate both the time and space required to develop new plant, animal, and food products. Finally, it may complement and extend

The knowledge and tools generated by molecular biology and biotechnology have stimulated a great deal of enthusiasm and redirected large sums of money in an effort to pursue knowledge in this area.

...several potential negative impacts of molecular biology on science may exist.

traditional methods used to enhance agricultural productivity and nutritional quality.

Still, several potential negative impacts of molecular biology on science may exist. Biotechnology and molecular biology continue and extend the basic methods and approaches of modern science. Their perspective, often called "logical positivism," attempts to reduce nature to small, definable pieces, subject to human manipulation and separated from broader questions of value. From this perspective, scientists control, measure, reduce, and divide nature in order to generate knowledge.

One concern is that this approach, while providing only partial knowledge, has become the dominant epistemology, often to the exclusion of other important alternative ways of knowing. As a consequence, whole plant and animal level research, such as traditional plant breeding, and systems level research programs such as agroecology, farming systems, and social assessments, which should be important complements to a comprehensive biotechnology research agenda, receive inadequate support. In U.S. colleges of agriculture, for example, employment opportunities in agricultural biotechnology and molecular genetics are growing, while employment in plant and animal breeding is declining. Between 1982 and 1988 full time equivalent (FTE) scientists conducting agricultural biotechnology research in the state agricultural experiment stations increased by 259 percent (273 to 682). In addition, staff working in this area rose from 472 in 1982 to 1131 in 1988. At the same time, overall FTE faculty positions in the state agricultural experiment stations increased by only 65 (National Association of State Universities and Land Grant Colleges, NASULGC, 1989). Therefore, much of the increase in the FTE's for biotechnology research represented a reallocation of existing positions. Interviews with state agricultural experiment station directors confirmed this conclusion. They indicated that many of these molecular biology positions had been created by reducing the scope of conventional breeding programs (Busch, *et al.*, 1991).

Another impact on science is the increase in the concentration of research funds and scientific talent at a small number of public and private institutions. For example, in the public sector, every U.S. state could afford and has had a conventional breeding program. Every state cannot afford and will not be able to have a comprehensive agricultural biotechnology program. Instrumentation and annual funding costs are particularly expensive, with start-up funds and operating costs two to three times that of other agricultural sciences (NASULGC, 1989). Concentration of public sector and scientific talent in a few states is already occurring. While 33 states are actively engaged in some promotion of biotechnology research and development, three states account for more than half the investment in biotechnology (Office of Technology Assessment, OTA, 1988). In agricultural biotechnology eight states account for over half of the state experiment station expenditures and nearly half of all science years for biotechnology research (NASULGC, 1989).

The new biotechnologies are also changing the division of labor between universities and industries with concomitant impacts.

The new biotechnologies are also changing the division of labor between universities and industries with concomitant impacts. While partnerships between universities and industries have existed for several decades, the new types of university and industry relationships in biotechnology (e.g., centers, institutes, research parks, public corporations) are more varied, more aggressive, and more experimental. They include: large grants and contracts between companies and universities in exchange for patent rights and exclusive licenses to discoveries; programs and centers, organized at major universities with industrial funds, that give participating private firms privileged access to university resources and a role in shaping research agendas; professors, particularly in the biomedical sciences, serving in extensive consulting capacities on scientific advisory boards or in managerial positions of biotechnology firms; and faculty receiving research funding from private corporations in which they hold significant equity. In a recent survey of biotechnology researchers at 40 major universities, 47 percent of these faculty consulted with outside corporations and eight percent held equity in a firm whose products or services were directly related to their own university research (Blumenthal, *et al.*, 1986).

The consequences of these collaborations may be both positive and negative. First, these university and industry collaborations may bring useful products to market more rapidly and may promote U.S. technological leadership in a changing world economy. This has been a major motivation behind a number of recent funding policies and laws of the federal government requiring this collaboration for federally funded research. Second, in light of funding stagnation at both the federal and state levels,

Even some scientists with public funding feel inhibited about discussing their work...

the collaborations are a means of raising new funds for university research. Third, these joint efforts may expand the scientific network, increasing communication between some industry and university scientists.

A number of concerns, however, have been raised regarding these new relationships. First, long term research, previously a major emphasis of the public sector, may decline. The private sector has short term proprietary goals, and, as a consequence, funding for research is also generally short term. A study in the mid-1980s of biotechnology firms revealed that nearly half funded research in universities. Of those funding research, 50 percent reported that projects spanned one year or less, while only 25 percent reported funding projects lasting more than two years (Blumenthal, *et al.*, 1986). In contrast, nearly all of the NIH extramurally funded programs and the USDA Hatch formula funded projects are for three years or more.

A second issue is a potential restriction of communication. Proprietary agendas have begun to inhibit the flow of information among biotechnology scientists and have raised concern about access to information. This is particularly true of university scientists with private sector grants, who often must delay public discussion of work, or its results, pending review by the sponsoring company. In one study 25 percent of industry sponsored biotechnology faculty, reported conducting research that belonged to the firm and could not be published without prior consent, while 40 percent reported that their collaboration resulted in unreasonable delays in publishing (Blumenthal, *et al.*, 1986). Even some scientists with public funding feel inhibited about discussing their work, for fear that some private company with the money, equipment, and time will utilize their ideas and perform the experimental work before they can. The net effect of these various developments appears to be a reduction of the free flow of information. Open communication is fundamental to public sector science, and indeed, one industry scientist remarked that more knowledge is generated by keeping an open environment for scientists (personal interview, 1987). Most breakthroughs do not come from just one laboratory; instead there is need for more information from a number of different laboratories. Communication among the scientists is crucial (OTA, 1986).

A third concern is a potential for conflict of interest or scientific misconduct. In interviews both public and private sector scientists stressed the potentially detrimental effects of granting private patents for work done in the public sector. These effects include potential favoritism, unwarranted financial advantage through privileged use of information or technology partly derived from publicly funded research, constraints on sharing of germplasm, and shelving of research which may be of interest to the public but not to the corporation (Lacy, *et al.*, 1988).

...the vast majority of small farms are now buffered from the effects of technological change, since the farm is no longer the primary source of income for their owners.

Recently, Derek Bok in his final President's report to Harvard University's Board of Overseers warned that the commercialization of universities may be the most severe threat facing higher education. Mr. Bok noted that as universities become "more entrepreneurial they appear less and less as charitable institutions seeking truth and serving students and more and more as huge commercial operations that differ from corporations only because there are no shareholders and no dividends." He concluded by saying that "it will take very strong leadership to keep the profit motive from gradually eroding the values on which the welfare and reputation of universities ultimately depend" (McMillen, 1991:A31).

IMPACTS ON FARMERS AND RURAL COMMUNITIES

The impact of agricultural change on rural communities is to a high degree proportional to the level of local dependence on agriculture. Today, nationwide, fewer than 40 congressional districts have more than 20 percent of their population living on farms (Sundquist and Molnar, 1991). The overwhelming majority of farms that once existed in the United States no longer exist and production is highly concentrated among the remaining farms. The largest 13 percent of farms now produce over 75 percent of the value of total production. In addition, the vast majority of small farms are now buffered from the effects of technological change, since the farm is no longer the primary source of income for their owners. Consequently, biotechnology will probably have less impact on the total number of farms than previous mechanical and chemical technologies adopted by farmers during the last 50 years. Moreover, it is likely biotechnology will not greatly exacerbate the decline in the number of farms, although it will certainly maintain present trends which indicate that farming will continue to be one of the fastest declining occupations in the next decade (Bureau of Labor Statistics estimates 28 percent decline between 1990-2000).

The extent of biotechnology's influence on the trend towards fewer and larger farms depends, in part, on how adoption affects the cost structure of farms. If biotechnology developments significantly alter costs, returns, competitive positions, and the special location of production, and if certain trade and farm policies are implemented, the potential impact of biotechnology could be relatively important. The Office of Technology Assessment (1986) has argued that these new technologies will be adopted by well financed, innovative farmers who are presumed to run the comparatively large farms. However, others have argued that biotechnology innovations will provide widespread benefits to the full range of farmers because new technologies will be used in traditional ways. Regardless of which farmers are likely to benefit, however, biotechnology will probably increase the value added off-farm at the expense of value added on-farm (Goodman, *et al.*, 1987).

Biotechnology may also accelerate the trend toward contract integration.

Other significant changes in the farming community may result if the information and products of this technology bypass the Extension Service and agricultural cooperatives. Previous products and information of biological research have been disseminated through the Extension Service. However, the development of new seed-chemical packages through biotechnology will emerge from private research, and public sector scientists will have limited knowledge with which to support extension programs. As a consequence, extension, and potentially agricultural cooperatives, may gradually be reduced to playing a secondary role in farm change. Moreover, many agriculturally based rural communities will continue the ongoing process of shrinkage and consolidation, as producers, and local supply and marketing firms continue to decline in numbers.

Biotechnology may also accelerate the trend toward contract integration, already common in the U.S., where commodities such as poultry and most processed vegetables are produced on contract. Such contracts specify the seeds, chemicals, planting and harvesting times and other aspects of farm production. These arrangements will further reduce the autonomy of farmers and will certainly reduce their contact with and need for extension services, agricultural cooperatives and local farm suppliers. The new biotechnologies may also restructure the relationship between farmers and researchers. Until very recently farmers were seen as the primary clientele of public sector research. However, the entry of molecular biology into agricultural research has increasingly been accompanied by the insertion of the agribusiness sector between farmers and researchers.

As a result it is quite possible that only problems of interest to the agribusiness sector will be the subject of public research agendas.

Another impact of biotechnology has been the stimulation of new moral and ethical debates regarding the limits of science.

IMPACT ON CONSUMERS

The new biotechnologies expand and extend researchers' ability to improve plants, animals, and microorganisms. For consumers this could mean dramatic improvements in the productivity and efficiency of food production and processing and the expansion and extension of food and nonfood uses of raw agricultural commodities. Consumers could benefit in the form of reduced prices, increased food safety and more nutritional foods. The new technologies also have the potential to change the very nature of food itself and to expand the range of possible food products. With molecular biology it is possible to move genetic material among plants and between plants, animals, and microorganisms. It is now possible to consider the production of new fabricated foods in which basic foods are broken down into their component parts (e.g., starch, fat, and sugar) and recombined into wholly new types of food. Such new forms of food may not be desirable to consumers and may make it far more difficult in the future for the consumer to determine the composition of the food and to maintain a balanced diet.

Another impact of biotechnology has been the stimulation of new moral and ethical debates regarding the limits of science. Public concern about a range of scientific developments including biotechnology, are resulting in a decline in public confidence in science and an increasing public perception of a likelihood of environmental risks from genetically-altered bacteria, plants and animals. The development of biotechnology is stimulating a wider range of concerns about science which extend beyond human health, environmental risk, food safety, and animal health issues and include such concerns as negative socioeconomic consequences and the morality of tampering with nature and life itself (Lacy, *et al.*, 1991).

IMPACT ON AGRIBUSINESS

John Hardinger, Director of Biotechnology at Dupont's Agricultural Products department, views biotechnology as a force to not only restructure farming but also to catalyze a major change in the structure of worldwide agribusiness. He notes that the application of molecular biology permits the various segments of the world's largest industrial sector to form logical linkages to other economic sectors that were never before practical. This 1.3 trillion dollar agribusiness sector (not counting feed and fiber) consists of the four basic elements: input suppliers, growers, processors and consumers. This system has experienced mechanical and chemical eras which contributed to increased productivity and efficiency. According to Hardinger the new biological and biotechnology era will further increase both efficiency and productivity, as well as provide the ability to change the quality of food and feed. Furthermore, it will lead to consolidation and new forms of vertical integration of the food industry (Busch, *et al.*, 1991).

The formation of new biotechnology companies increased dramatically from 1979 to 1983, with more than 250 small venture capital biotechnology firms founded in the U.S. Proliferation of these risk-taking companies helped raise billions of dollars from private investors and gave the U.S. a competitive lead in the early stages of biotechnology commercialization. By the late 1980s the number of these firms had grown to over 600. However, consolidation has begun in the industry with mergers, bankruptcies and major multinational corporation investments. Indeed, 80 percent of the funds in venture-capital firms have been invested in just ten companies (OTA, 1988).

In the early 1980s, multi-national corporations began to recognize the potential of biotechnology and to develop their own research and development capacities. These corporations began diversifying into every field or specialty that used living organisms as a means of production. The new biotechnologies appeared to further reduce the distinctions among the traditional industrial sectors, rendering corporate boundaries virtually unlimited. Those large multinational corporations specializing in oil, chemicals, food, and pharmaceuticals have taken the leadership in agricultural biotechnology research and development (e.g., American Cyanamid, Campbell Soup, Ciba Geigy, Dupont, Eli Lily, Monsanto, Rhone-Poulenc, R. J. R. Nabisco, Shell, Sandoz and Standard Oil). These corporations have also established research contracts, joint projects, exclusive licensing and marketing, equity positions, and control or ownership in the venture capital firms. By the late 1980s a small number of large multi-national corporations had significant positions in all the major biotechnology firms and provided over half of total funds being invested in developing this new technology (Busch, *et al.*, 1991).

This concentration, accompanied by horizontal and vertical integration across industrial sectors, reflects the mergers, acquisitions and concentration in the food processing industries as traditionally nonfood industries dramatically expand their investments. This trend is also apparent in the input industries. Of the top seven pesticide corporations, five ranked among the world's largest twenty seed companies with only Bayer and Dupont having marginal seed interests. Moreover, of the ten top seed companies, eight have significant interest in crop chemicals. Most analysts predict biotechnology will continue and accelerate this trend towards increasing concentration of power in the hands of a small number of large multinational corporations.

Consequently, development and commercial control will be in the hands of corporations that transcend geographic boundaries and hold limited national allegiance.

Consequently, development and commercial control will be in the hands of corporations that transcend geographic boundaries and hold limited national allegiance. Within this context, people question how we can ensure that democratic participation will occur in the decision-making processes surrounding the development and commercialization of biotechnology. This is difficult within national boundaries and generally prohibitive internationally, given current governmental structures.

INTERNATIONAL IMPACTS

The new technologies offer the hope of increasing crop yields where population growth is outstripping the food supply. In a recent parliament meeting in India, biotechnology was acknowledged as the lifeline for the whole of Indian agriculture, offering opportunities for increased sustainability, profitability and international competitiveness (Ministry of Foreign Affairs, The Netherlands and the University of Amsterdam, 1990). Other nations have been equally optimistic regarding the prospects of agricultural biotechnology (Deo, *et al.*, 1989).

It has been proposed that the direct use of molecular biology in conjunction with plant propagation and breeding could dramatically increase crop productivity and overall food production in developing countries. Tissue culture techniques are already creating more drought and disease resistant varieties of cassava, oil palms, and groundnuts. Embryo transfer may raise the reproductive capacity of livestock. In Africa genetically engi-

Whose needs and goals will be served and whose neglected are perhaps the most important agricultural and social questions of the coming decade.

neered vaccines and drugs may cure fatal or debilitating diseases (e.g., onchoceriasis-river blindness, Guinea worm disease, schistosomiasis and trypanosomiasis) thus opening up new agricultural and grazing areas (Gibbons, 1990; Barker and Plucknett, 1991).

Yet despite biotechnology's great promise for feeding the world's rapidly growing population, particularly in developing nations, science and policy-makers admit it will not be easy to ensure that this technology has the desired positive effects. First, there is legitimate concern that the developed nations will use their technology to undercut traditional Third World exports, such as sugar, vanilla, cocoa butter, and other important cash crops. Genetic engineering is already being applied to bacteria, yeast and fungi to produce starter cultures with specific metabolic capabilities in food fermentation. These processes, combined with the new cell culture techniques, are being used to transform the production of certain agricultural commodities into industrial processes. In principle, any commodity that is consumed in an undifferentiated or highly processed form could be produced in this manner and product substitutions could be easily introduced. Similarly, although with greater difficulty, tissue culture techniques could be used to produce edible plant parts *in vitro*. In short, agricultural production in the field could be supplanted by cell and tissue culture factories (Busch, *et al.*, 1991).

Several companies are now capable of phytoproduction of a natural vanilla product in the laboratory. A genetic modification of oilseed plants to convert cheap oils (e.g. palm or soybean oil) into high quality cocoa butter is well advanced. Biotechnology is also being used to produce substitutes for sugar as an industrial sweetener. Several major corporations in the U.S. and Europe, (e.g., Unilever and Ingene) are attempting to use recombinant DNA technology to produce the thaumatin protein, one of the sweetest known substances. Successful development of a thaumatin product through genetic engineering may continue a transition to alternative sweeteners, eliminating the market for beet and cane sugar and capturing the valuable sweetener market. Even moderate success in realizing these product substitutions would have profound effects around the world, most immediate and important would be the restructuring of global markets (Persley, 1990).

Another issue focuses on environmental risks. Because the environmental release of genetically modified organisms may have hazardous effects on the ecosystem, in many of the industrialized countries the public has pressed for the adoption of safety regulations. These regulations, however, may restrict biotechnology experiments. As a consequence, researchers and biotechnology companies are attempting to relocate their experiments to countries with limited or no safety regulations. This may result in the movement of possibly hazardous biotechnology experiments to Third World countries.

...agricultural biotechnology may shift the geographic location of agricultural production from one Third World country to another or from the Third World to the First World.

A further concern is that biotechnology will increase the disparities between the developed and developing nations. With the shift in applied research and associated product development from the public to the private sector, the benefits from the new biotechnologies may become less widely available. Moreover, the products developed are unlikely to be ones which are important to the poor developing countries, particularly in the tropics. Biotechnology research has emphasized temperate zone animal reproduction, breeding, veterinary health care and animal nutrition, and temperate zone plant improvement. Little or no work is currently being directed at transformation of tropical crops important to developing countries. This could further widen the gap between the agriculture production methods in the North and the less advanced practices in the South (Deo, *et al.*, 1989).

The Third World might be able to counter these technological developments by enhancing its own scientific capabilities. But this is unlikely to occur. Many developing countries have no basic research capacity, limited capabilities to adapt biotechnological advances to local conditions, and few resources to attract transnational corporations.

In conclusion, agricultural biotechnology may shift the geographic location of agricultural production from one Third World country to another or from the Third World to the First World. For many Third World countries, dependent on one or two agricultural commodities for their continued viability, this production and market restructuring and increased productivity gaps could result in a collapse in existing markets. Significant numbers of farmers and farm workers could find themselves with no products to sell. This could increase the already high Third World debt and exacerbate the deficit in balance of payments in Third World

The ultimate direction this technology takes will be determined by the actors who participate in its development.

countries. Political instability, already a problem in the developing world, would doubtless increase.

The effects of these possible changes in the patterns of world trade are also likely to be felt in the West. With developing countries experiencing economic deterioration they could no longer serve as a main market for developed country exports, creating economic and social stress in the developed nations as well. For the continued well-being of an increasingly global economy, a conscientious effort must be made to help developing nations acquire the appropriate technology, establish and maintain an infrastructure for support of applied research, and improve their capacity to evaluate this new technology in terms of their own public good.

CONCLUSION

Addressing issues relative to the new biotechnologies at this early stage offer us an opportunity to assess these technologies before they actually exist and to examine the alternative paths for their development. Although changes and developments are proceeding rapidly, they are neither inevitable nor totally shaped. As Winner (1986:29) observed "By far the greatest latitude of choice exists the very first time a particular instrument, system or technique is introduced." The ultimate direction this technology takes will be determined by the actors who participate in its development.

This paper has discussed a number of both positive and negative social and economic impacts biotechnology may have on science, farms and rural communities, consumers, agribusiness, global markets and the international community. We suggest that biotechnology may increase inequities not only among various groups in our society, but also between developed and developing countries and among the more and less advanced developing countries. These potential impacts raise complex ethical and policy questions. A more careful review and monitoring of the scientific developments is essential, and a more detailed assessment of the fourth criterion, the social and economic impacts of particular technological changes, is needed. Further, we need to incorporate the fourth criterion into our decision-making and develop policies and long-range planning capacities to address the potential scenarios. Finally, we need to balance research programs, nationally and globally, in terms of time frame, proprietary nature and level of analysis (molecular, cellular, species and system) to ensure an agenda that is environmentally sound, enhances our health and focuses on building sustainability in both our fields and communities. Whose needs and goals will be served and whose neglected are perhaps the most important agricultural and social questions of the coming decade.

BIBLIOGRAPHY

Barker, R. and D. Plucknett. 1991. Agricultural biotechnology: A global perspective. B. R. Baumgardt and M. A. Martin eds.. *Agricultural Biotechnology: Issues and Choices.* West Lafayette: Purdue University. pp. 107-120.

Blumenthal, D. M., M. Gluck, K. S. Louis, M. A. Stoto, and D. Wise. 1986. University-industry research relations in biotechnology: Implications for the university. *Science* 232: 1361-6.

Busch, L,. W. B. Lacy, J. Burkhardt and L. R. Lacy. 1991. *Plants, Power and Profit: Social, Economic, and Ethical Consequences of the New Biotechnologies.* Cambridge, MA and Oxford, England: Basil Blackwell.

Busch, L. and W. B. Lacy. 1991. Biotechnology: Prospects and economic, social, and ethical consequences for agricultural and alimentary systems. Testimony at the parliamentary conference: *Agricultural and Alimentary Systems on a Global Scale.* Rome, Italy.

Deo. S. D., J. S. Silva, L. Busch, W. Lacy and N. Mohseni. 1989. Agricultural biotechnology in India and Brazil: Creating new technological dependencies: *International Journal of Contemporary Sociology* 26(3-4): 217-233.

Gibbons, A. 1990. Biotechnology takes root in the Third World. *Science* 248:962-963.

Goodman, D., B. Sorj and J. Wilkinson. 1987. *From Farming to Biotechnology: A Theory of Agro-Industrial Development.* Oxford: Basil Blackwell.

Harlander, S. K., J. N. BeMiller and L. Steensen. 1991. Impact of biotechnology on food and nonfood uses of agricultural products. B. R. Baumgardt and M. A. Martin eds. In *Agricultural Biotechnology: Issues and Choices.* West Lafayette: Purdue University, pp. 41-54.

Lacy. W., L. Lacy and L. Busch. 1988. Agricultural biotechnology research: Practices, consequences, and policy recommendations. *Agriculture and Human Values* 5 (3): 3-14.

Lacy, W., L. Busch and L. Lacy. 1991. Public perceptions of agricultural biotechnology. B. R. Baumgardt and M. A. Martin eds. In *Agricultural Biotechnology: Issues and Choices.* West Lafayette: Purdue University, pp. 139-162.

McMillen, L. 1991. Quest for profit may damage basic values of universities, Harvard's Bok Warns. *The Chronicle of Higher Education.* XXXVII (32): A1, A31.

Metheny, B. and J. Monahan. 1991. President's FY92 budget proposes $4.1 billion for biotechnology research and development. *Genetic Engineering News* March:3.

Ministry of Foreign Affairs and the University of Amsterdam, The Netherlands. 1990. Biotechnology in India. *Biotechnology and Development Monitor.* 5 (December): 10-14.

National Association of State Universities and Land-Grant Colleges (NASULGC). 1989. *Emerging Biotechnologies in Agriculture: Issues and Policies.* Progress Report VIII. Washington, DC: NASULGC, Division of Agriculture, Committee on Biotechnology.

Office of Technology Assessment (OTA), US Congress. 1988. *New Developments in Biotechnology: US Investments in Biotechnology.* OTA-BA-360. Washington, DC: US Government Printing Office.

____. 1986. *Technology, Public Policy and the Changing Structive of American Agriculture.* OTA-F-285. Washington, DC: US Government Printing Office.

Persley, G. J. 1990. *Beyond Mendel's Garden: Biotechnology in the Service of World Agriculture.* Biotechnology in Agriculture Series, No. 1. Oxon, United Kingdom: CAB International.

Sundquist, W. B. and J. J. Molnar. 1991. Emerging Biotechnologies: Impact On Producers, Related Businesses, and Rural Communities. B. R. Baumgardt and M. A. Martin eds. In *Agricultural Biotechnology: Issues and Choices.* West Lafayette: Purdue University, pp. 23-40.

Weber, G. 1990. USDA–ES. Extension biotechnology coordinators memo. Washington, D.C., March 24:1–2.

Winner, L. 1986. *The Whale and the Reactor.* Chicago: University of Chicago Press.

CONSUMER CONCERNS
*GIVE US ALL THE DATA**

I am not going to focus on public concerns about all of agricultural biotechnology but primarily on bovine growth hormone or bovine somatotropin (BST). I do this because BST is going to be the first major product in agricultural biotechnology to come out—if it is approved. There has been a lot of controversy surrounding this product, and I think there is a lot of consumer concern about BST.

MICHAEL HANSEN

Consumers Union
Consumers Policy Institute
101 Truman Avenue
Yonkers, NY 10703-1087

Active in work related to both biotechnology and pesticides, Hansen is a member of the biotechnoloy working group and has written several books on biotechnology.

I would like to raise a couple of general issues and then spend time discussing a number of internal documents from the Food and Drug Administration (FDA)—letters and memoranda sent to companies that paints a very different picture of some of the problems with BST than what has come out publicly.

CONSUMER PERCEPTIONS OF BIOTECHNOLOGY

A number of studies have shown that consumers are very wary about new technologies and in fact, are very wary about biotechnology. In a survey that was done for the National Dairy Board in 1986, one of the conclusions was "that the strongest impression in the consumers' mind is that most problems are result of man's [sic] interference with nature, a tampering with the natural order of things for the sake of profit or convenience or both." Some of the people who have criticized the media and criticized the critics of biotechnology have tended to say that the problem with the general public is that they are very uninformed. If only they were scientifically more literate, they would understand this technology and come to accept it. I think that consumers are not as ignorant as a lot of people might think.

A study done for the National Dairy Board in 1990 asked consumers about their awareness of a number of issues and how important they felt that issue was (i.e., whether they were concerned with it or not). Two

*Edited from tape of lecture. The letters referred to were shown in slides.

...consumers are not as ignorant as a lot of people might think.

things can be seen. A large percentage of consumers were aware of both the Alar® contamination of apples and the recall of grapes from Chile because of cyanide contamination—83 and 85 percent respectively. If you look at the number of consumers, of the percent that were *not* concerned about those issues, among those that were aware, you find it is fairly significant—27 percent and 29 percent. When you look at other issues which I think are more important than either the Alar® scare or the recall of grapes from Chile, you see that for Salmonella in fresh poultry, drugs given to cattle, the use of hormones to increase milk in dairy cows or the antibiotics used to treat livestock that are harmful to humans—there is a smaller percent awareness of those issues in general. Yet the percentage of consumers who do *not* think that issue is a problem (if they know about it) is much smaller. What this shows is that the consumers are a little bit more aware than some people give them credit for. Of those that were aware of the Alar® controversy and the recall of grapes from Chile, more than 2½ times those people were not concerned relative to concerns over drugs given to cattle, the use of the bovine growth hormone and antibiotic questions.

WHO DO CONSUMERS TRUST?

Let us now look at who consumers tend to trust. A survey in *Dairy Today* in 1990 asked, "Who would you trust regarding BST safety?" Eighty-four percent said that they would trust consumer safety groups, 54 percent university researchers, 45 percent the American Medical Association and 34 percent the FDA. The family doctor is the number one person that consumers trust for sources of information about the safety of food. *Consumer Reports* is number two. The drug companies are at the bottom of the list. It seems as though consumers are not very trustful of the companies. They also are not trustful of FDA. Is there is a reason for that? I think that indeed there might be. In 1990 the National Dairy Board asked consumers questions after they were given information in the form of statements on BST. Only 25 percent of the consumers agreed with the statement "milk from cows treated with BST is completely safe"; 83 percent agreed with the statement "the long-term or the long run health implications of BST are not known"; only 29 percent agreed with the statement "testing has shown that the milk from cows treated with BST is completely safe"; and 63 percent agreed with the statement "milk from the cows treated from BST may be harmful to humans." This survey shows there is a lot of concern out there among consumers.

...we have to ask, is this concern legitimate or is it based on not much information?

BST AND SECONDARY DRUG RESIDUES

Now, we have to ask, is this concern legitimate or is it based on not much information? If you look at the studies which have come out they have tended to say that there is not a problem with BST (i.e., with the consuming of milk and meat products from the experimental test herds). In fact, in 1986, when the FDA made the decision that there were no human health impacts and therefore permitted the marketing of milk from these experimental test animals, there was a lot of controversy. Since then there has been an article published in *Science* magazine and most recently, an Office of Technology Assessment (OTA) report came out which basically said that there is no concern for human health. Concerns are my work in this area—the main problem I have with FDA's position is the one thing they have missed—a primary human health concern with the potential for the secondary drug residues in milk. (I also have some concerns about insulin-like growth factor, but will not discuss them here.) The concern about drug residues arises when the animals treated with BST and if there is an increase in mastitis and other bacterial infections, then that could lead to increased use of antibiotics, which in turn leads to antibiotic residues in milk. This is a very important issue.

I have been told, off the record, by people from the dairy industry that they are concerned about the unregulated drug use in animal agriculture.

The Government Accounting Office (GAO) late 1990 released a report which says that FDA cannot guarantee the safety of the milk supply, and that their testing procedures are highly flawed and do not pick up a lot of antibiotics. I have been told, off the record, by people from the dairy industry that they are concerned about the unregulated drug use in animal agriculture. The FDA, also off the record, admitted that there is a concern there. So our position and concern at Consumers Union is that if this technology can be shown to increase disease rates in these cattle, then that in turn potentially can lead to greater contamination in milk with antibiotic residues. If that is a significant human health concern, then how can FDA say that there are no human health effects when they have not explicitly looked at that issue? Further, the FDA has explicitly not yet discussed whether there are any health impacts on the cows.

HEALTH IMPACT ON TEST HERDS

Quoting now from the OTA report, which came out this spring, on the effects of bovine health and stress, "catastrophic effects such as the instance of ketosis (which is under production of glucose) fatty liver, crippling lameness, milk fever (which is feverish disorder following parturition), mastitis (inflammation of the udder), sickness, suffering and death have all been postulated to occur with BST." This is the important part—"however, no such effects have been observed with BST supplementation of dairy cows in any scientifically valid published study, nor have subtler health effects been in evidence." So what OTA is saying is that there is no evidence whatsoever of problems such as mastitis. Now, I tend to think that there is some evidence of such problems. I base this belief on internal documents from the FDA and other papers.

This indicates that the FDA internally knew that things like environmental assessments and health effects are not legitimate confidential business information, but was going to make people sue them to get the information out.

In a memorandum of a meeting between representatives from Monsanto and representatives from the Center for Veterinary Medicine, the first line says, "Mr. Matheson requested this meeting in order to formally request the preparation of an environmental assessment for this and all other INADs." An INAD is an Investigational New Animal Drug. An INAD had been filed by Monsanto which involved the use of recombinant DNA technology in the production of a new animal product. That new product is bovine growth hormone. By law, these environmental assessments should be done *before* an INAD is granted. The first INAD for Monsanto was granted around 1983. Did they prepare an environmental assessment? No.

The second page of this memorandum addresses the petition submitted by the Foundation on Economic Trends. In March of 1986, the Foundation on Economic Trends and sixteen other groups signed onto a petition that demanded that FDA do an environmental impact assessment on bovine growth hormone. FDA turned down their petition saying that an environmental assessment and an environmental impact statement had already been prepared. The groups then petitioned, using a freedom of information act, to get copies of the environmental assessments. FDA said, "sorry, that's trade secrets, we cannot release this." The Foundation took them to court, 2½ years later, the documents were released. FDA had told the company that the environmental assessment and CMDs findings would remain in the INAD unless disclosure was required through court proceedings. This indicates that FDA internally knew that things like environmental assessments and health effects are not legitimate confidential business information, but was going to make people sue them to get the information out.

Now I am going to go on to deal with two sets of letters. They are from FDA, the first set to Eli Lilly and the second set to Monsanto. These are what are called New Animal Drug Application (NADA) incomplete letters. The way the regulatory system works, you get an INAD permit which permits you to ship a drug across state lines and gather data in farmers fields and at universities. After that data is gathered and submitted to the agency, the company then submits a NADA asking the agency to let them sell this product and to make a given label claim. In Eli Lilly's letter of October 8, 1987, they requested approval to market a sustained release subcutaneous injectable BST with the claim of increasing milk production and improved feed efficiency. The FDA told them that the submission was incomplete, making specific comments. I want to highlight a few things in this letter that caused us to be very upset. The first thing has to do with the efficacy. In this letter, FDA talked about a serious error in the milking pattern in the Canadian trial. Critics of this research have talked about how there has been manipulation of data to show that there is not any problem or to show that the drug works. In this Canadian trial they were establishing what the optimal milking interval was for the highest producing cow. They used that milking interval with lower producing cows although it was suboptimal for them and could be detrimental to the milk yield. What the researchers said is that the cows were continually moved such that the top producers were exposed to a near ideal milking interval, while the low producers' milk yields were probably diminished due to a wide milking interval. Logically, control animals would, on average, end up with the lower producers, particularly by the end of the lactation, while BST treated cows would be among the top producers. The continual movement of cows throughout the length of the study prevented any possibility of factoring out this problem. Thus there was a bias in the way that the study that was done.

A French study commented on by FDA looked at the effect of BST on milk composition. It demonstrated no alteration or no effect on the fatty

It is interesting that this negative result never appeared in the literature.

acid content or the protein content of milk. However, FDA had to point out to Eli Lilly that this data was not very useful because just the milk yield was not changed by treatment and there were no differences between the treated cows and the experimental cows. It is not feasible to extrapolate those milk composition results to a situation where milk production was increased by BST. Again, a flawed study. The summary of sensitization studies stated that the product (milk) elicited a positive sensitization response. The same thing was brought up in the Monsanto letter. In the animal health and reproduction target animal safety section of the FDA letter to Eli Lilly, it says point blank, "We are concerned about the increased incidents and duration of mastitis and reproductive parameters in the field studies."

Now let us turn to the Monsanto letter. Again, Monsanto was trying to get FDA approval. They sent in all their data asking for approval to market this product (BST) and to make label claims of increased milk production and increased feed efficiency. All of their data that was sent in says because no statistical differences were found between treatment groups for feed efficiency, they did not bother to review the data further. That is very curious. All the data that Monsanto had sent in, as of late 1987, could not demonstrate any differences in feed efficiency. Yet if you look in the literature, there are all these studies which demonstrate increases in feed efficiency. It is interesting that this negative result never appeared in the literature.

When we go to the target animal safety section of the Monsanto letter, FDA discusses the problems with mastitis—in 3C, it says "you should address the use of gentomycin and tetracycline, which are not approved for the treatment of mastitis in dairy cattle." Critics have pointed out that internal documents leaked from Monsanto indicated that there was use of drugs which were not permitted in dairy cattle. The letter continues, "the overall conclusion for the mastitis was data presented indicates that there was an increase in mastitis at levels at which you wish to market bovine somatotropin". Responding to the reproductive data FDA says that you have compromised the usefulness of your reproductive data by the use of prostaglandins and progesterone assays. It is not possible to evaluate the effect of BST on reproduction if concurrent use of reproductive hormones in diagnostics tests masks or otherwise alters the effect of the drug. In 4B of this letter it talks about the data indicating that the regimen outlined in their protocols was not adhered to and in 4C it indicates that the use of prostaglandins varied in time, frequency and reasons for use, and were used for unapproved claims such as metritis. "Some cows", the letter said, "received a drug that was not approved for use in dairy cattle." The overall conclusion was that "your data indicates that there are reproductive problems at the doses at which you wish to market bovine somatotropin."

Turning to injection site syndrome, FDA said that the data indicated that the severity of injection site response increases with time. In 5C it continued, "your data indicate that there may be an unacceptable level of tissue reaction with prolonged use of BST in the range of doses for which you wish approval. Lameness and displaced abomasum may be drug related. You (Monsanto) have not established a margin of safety, nor have you established a no effect level for some of the parameters in your submission." Based on available data, this is particularly true of major clinical entities such as mastitis and reproduction. Remember, FDA is saying that Monsanto could not demonstrate a "no effect" level. The letter concludes with "if you seek approval of a range of 250 to 500 milligrams in cows or heifers, you may not even have a 1–fold margin of safety. Under current standards, this is unacceptable for an over-the-counter approval." On the last page of the letter, FDA talks about mitigation measures. They say "you should describe steps you (Monsanto) are taking, including labeling, to present sensitization by those occupationally exposed to the BST protein."

So, both of these letters, one to Eli Lilly and one to Monsanto, indicate that there seem to be problems. In fact, two weeks after the latter letter was sent, there was a meeting between Monsanto and the FDA to talk about the target animal safety portion of this NADA incomplete letter. Dr. Lehman, head of the production division of the drug section, very clearly expressed concern over the increase in days that cows were affected by mastitis during the BST treatment. Again noting that there was mastitis. Concern was also raised over injection site swelling. The use of reproductive agents was discussed. Some drugs may be used for clinical conditions; however, they also are used to get cows pregnant to cover up reproductive problems. There was a lot of concern there. About a year later, there was a memo written by Dr. Guest after meeting with the National Milk Producers Federation. In it, they are talking about how to manage the consumer

Consumers have the right to know about health impacts.

perception. They ask that in view of the likely problems with cattle safety and the need for careful management of the cattle, should the product be prescriptioned? It was also asked whether prescriptioning BST might also give some sense of security to the consumer. My concern with a lot of these documents, and I think Consumers Union's too, is how can the OTA say that there is no evidence whatsoever that there are problems such as mastitis with BST use when, internally, it seems like FDA is saying that there are problems with mastitis and with reproduction. Unfortunately none of this data was made public.

Now, those letters that I discussed (shown as slides at the meeting) *are* three years old. Maybe I am willing to admit that companies have been able to overcome these problems discussed above in the last three years. But the problem I still have is why is it that none of the negative effects discussed in these letters have never made it into the literature? And if the companies and the FDA want to tell us that there is not any problem, why should we trust them?

GAINING TRUST

What I think needs to happen, if you want consumers to have some trust in what you are doing, is the data has to be made readily available. Consumers have the right to know about health impacts. I think that the data on the adverse effects of BST not being made public while all the positive data is being made public. This is of a lot of the public's concern. We would like to see all the health and safety data released to the public. I would just like to say to those other companies working on applications of biotechnology—make your data public, try to be up front with the critics.

WORKSHOP WHITE PAPERS

c. 1991

HERBICIDE TOLERANCE IN CROPS

WALTER R. FEHR

Director of Biotechnology
Iowa State University
1010 Agronomy
Ames, IA 50011

As Director of Biotechnology Fehr oversees teaching, research and extension activities. Involved with graduate education and research in plant breeding, his specialty is soybean breeding and genetics.

PART I

Herbicide tolerance in crop plants is not a new concept. When I grew up on the farm in northern Minnesota during the 1950s, I helped my Dad spray the wheat and barley with 2,4–D to control wild mustard. The wheat and barley were tolerant to the 2,4–D. We also sprayed the farmyard with 2,4–D to control dandelions and other broadleaf weeds. The Kentucky bluegrass and bromegrass were tolerant to 2,4–D. We could not use the 2,4–D to control wild mustard in our alfalfa or sweet clover because the crops would have been severely damaged by the herbicide due to their lack of tolerance.

If my father understood herbicide tolerance in the 1950s, why are we discussing it in the 1990s? We are discussing it today because of our ability through biotechnology to make crops tolerant to herbicides that previously would have caused them severe damage. This new ability has raised questions about herbicide tolerance relative to its impact on human safety, environmental safety and the social structure of agriculture. I would like to briefly discuss each of those topics, but first let me review the science that has generated the questions.

Before biotechnology entered the scene, new herbicides were developed by generating an array of new molecules in the laboratory and spraying them on weeds in the greenhouse to determine their effectiveness. If a molecule damaged the weeds, it ultimately was sprayed on crops to determine their level of tolerance. If the crop could not tolerate the herbicide at rates that were needed to damage important weeds, there was no practical way to alter the crop to give it an acceptable level of tolerance, unless there were genetic differences among cultivars within the species. Occasionally, such genetic differences were found. In soybean, tolerance to metribuzin varied among cultivars. Farmers who used the herbicide had to select cultivars that had an adequate level of tolerance. Some soybean breeders selected for metribuzin tolerance as part of their cultivar development programs.

I suspect that most people never knew that some soybean cultivars were bred for tolerance to metribuzin, at least I do not recall any national conferences at which the practice was debated. So why are we discussing it today? I believe it is an issue today because biotechnology is involved in generating new genes for herbicide tolerance and new methods for transferring the genes among species. I know that some people believe strongly that herbicide tolerant crops (HTCs) developed by the new methods should be called genetically engineered HTCs (GEHTC) or some other designation to distinguish them from the herbicide tolerance that is naturally available in a crop. In the interest of space, however, I will use HTC.

Several methods are being used to develop herbicide tolerance in crop plants. The most simple method is artificial mutagenesis. DuPont developed tolerance to sulfonylurea, sold commercially as Pinnacle®, in soybean by treating seeds of soybean cultivars with chemical mutagenesis. The treated seeds were planted, the resultant plants were naturally self-pollinated, and the self-pollinated seeds were harvested. The seed were sown and the plants were exposed to chlorsulfuron. A mutant plant was found that tolerated the herbicide. The gene conferring the resistance is now being put into high-yielding cultivars by traditional plant breeding methods.

Tissue culture was used by scientists at the University of Minnesota to develop tolerance in corn to the herbicide sethoxydim, which is sold commercially as Poast®. Cells of corn were grown in a media that contained the herbicide. Cells that survived were regenerated into whole new plants, and the regenerated plants and their progeny had tolerance to the herbicide. The gene conferring tolerance to sethoxydim is being transferred into commercial hybrids by traditional plant breeding methods.

A third approach has been used by Monsanto to develop tolerance to glyphosate, which is sold commercially as RoundUp®. They identified a gene in petunia that controlled tolerance to the herbicide. They isolated the gene from petunia and transferred it to soybean by use of *Agrobacterium tumefaciens*. No traditional plant breeding has been used in the Monsanto approach.

Regardless of the method used to develop tolerance to a herbicide, the issues raised about the technology are essentially the same. In fact, past research by soybean breeders in selecting for metribuzin resistance that largely went unnoticed would probably be called into question today.

We are now ready to examine the issues that are being raised about HTCs. For the following discussion, I will rely heavily on the discussion at a conference on HTCs held at Iowa State University (ISU) in October 1990. The report of the proceedings of the conference is published in its entirety in part II of this paper (see page 185). I intend to address only a few of the issues discussed at the conference.

The safety of HTCs for human and livestock consumption has been questioned. If tolerance is extremely high and farmers apply a herbicide at unusually heavy rates, the crop may not be able to completely degrade the compound and a portion of it may end up in the part harvested for feed or food.

There does not seem to be any debate about the importance of determining the safety of HTCs for feed or food consumption. Approval of the Environmental Protection Agency (EPA) must be obtained before a herbicide is registered for use on a crop. This approval will require an evaluation of the presence of the herbicide or any undesirable toxicants.

As a precaution against excessive application of a herbicide, it was recommended at the ISU conference that herbicide tolerance in a crop should be designed to withstand only several times the normal rate needed for weed control. If a farmer used the herbicide at such a high rate that the plant could not break it down, crop injury would occur.

The influence of HTC on environmental safety has been questioned. After listening to many discussions about environmental safety, I have concluded that the concerns are not specific to HTCs, but to herbicides in general. The issues raised about environmental safety also seem to be difficult to separate in most discussions from socioeconomic questions about general herbicide use. Those who oppose HTCs based on environmental concerns are generally the same persons who oppose the use of any herbicide because of its potential influence on the social structure of agriculture.

I would like to separate environmental safety and socioeonomic concerns for a moment because most farmers I know are concerned about the environment, but do not oppose the use of herbicides for socioeconomic reasons. The use of HTCs will be negative for the environment if they do not result in development of more environmentally safe compounds that can be applied at minimal rates. There are several reasons why I am optimistic that more environmentally safe compounds will be developed through the use of HTCs. First, all chemical companies are keenly aware of the demand by society for safe air, water and food. Second, the EPA is keenly aware of the demand by society for safe air, water and food. Third, state governments are keenly aware of the demand by society for safe air, water and food. This awareness is also shared by farmers, food processors, food manufacturers and food merchandisers.

I am also optimistic that the new compounds used with HTCs will be applied at minimal rates. As a person who has spent his entire life surrounded by farmers, farm magazines and other means of agricultural communication, I have difficulty understanding those persons who equate HTCs with herbicide misuse. I can tell as many stories as anyone else about the misuse of herbicides. I am concerned when farmers, homeowners and other users of chemicals do not treat products with adequate respect for their own personal health and safety. But when it comes to rates of application, I believe the value of the dollar has been and will continue to be the main deterrent to excessive application. I do not know a single farmer who does not consider cost when choosing a herbicide. One of the most important ways to reduce herbicide cost is to reduce the rate per acre. I cannot count the number of articles I have read this winter in farm magazines about banding of herbicides to reduce the cost of weed control. So you must excuse me for not understanding those persons who argue that HTCs and excessive herbicide use go hand-in-hand.

The socioeconomic impact of HTCs is difficult for me to evaluate. Those who express the greatest concern about the negative impacts of HTCs on the social structure of agriculture are generally the same persons who have a clearer picture than I do of what the social structure of agriculture should be in the future. It is argued that HTCs will permit farmers to achieve better weed control more easily, which will permit them to increase their farm size causing a reduction in the number of farmers. My problem with that argument is that weed control is not the factor that determines farm size in the Midwest. Availability of family labor, capital and land itself seems to be far more important than weed control in determining farm size. I think about the conversations I heard this winter concerning land that was sold in Iowa. Weed control was never mentioned as a reason for buying or not buying the land.

A central issue for U.S. agriculture that pertains to HTCs and other products of biotechnology is the appropriate social structure of agriculture. I have suggested that in the Midwest, farm size will not likely be determined by the presence or absence of HTCs. But what if you believe that HTCs will result in large farms, should that be sufficient reason for prohibiting the sale of HTCs? On whose vision of the social structure of agriculture will we base that decision? Should we use the vision of the National Farmer's Organization, the American Farm Bureau, the American Soybean Association, or a multitude or other farm organization and special interest groups? When I ask the question of vision to farmers in Iowa, the answer is clear. "If a product is considered safe by the federal regulatory agencies, I will decide if it makes economic sense for my operation. I don't need anybody else making my economic decisions."

If we assume that HTCs will be developed and sold in the United States, will they be adopted by farmers, and what precautions, if any, should be taken when they are used? The adoption by farmers will be decided on the basis of the suitability of current herbicides for weed control, the performance of the new herbicide that can only be used with the HTC, and the performance of the HTC cultivars with respect to yield, pest resistance, composition and other important economic traits. It is impossible to generalize about the suitability of current herbicides. Most weeds are adequately controlled through a combination of seedbed preparation, herbicides, mechanical cultivation and hand weeding. At the ISU conference, it was estimated that only five percent of the acreage in Iowa has a weed problem that current herbicides cannot adequately address.

Performance of HTC cultivars will be a major consideration in their adoption. There is no reason for assuming that HTC cultivars with competitive performance cannot be developed. However, adding herbicide tolerance as a selection criterion in a breeding problem will involve more cost for cultivar development. Unless all breeders of a crop insist on having herbicide tolerance in all cultivars, farmers will have a choice between tolerant and nontolerant ones. If HTCs are not competitive, farmers will discriminate against them.

Precautions in using HTCs include factors that relate to all herbicides. To minimize the risk that herbicide–tolerant plants will evolve when a herbicide is used, compounds with different modes of action should be rotated. There is no question that differences in herbicide tolerance among crops has necessitated the rotation of some products. Atrazine was suitable for corn, but not soybean. Treflan®(trifluralin) was suitable for soybean, but not corn. However, Lasso®(alachlor) could be used for both corn and soybeans. Did farmers use Lasso continuously, or did they rotate different herbicides? They generally rotated herbicides because other suitable products were available. If no suitable alternative was available, they probably would have gambled and used the same product on both crops. Therefore, rotation of products with HTCs will probably be determined by the alternatives that are available. The worst thing that could happen would be a reduction by the private sector in their research for the identification of new compounds, with or without the necessity for HTCs.

A second precaution with HTCs is to avoid growing them in areas where they can naturally cross with weed species. If a gene conferring tolerance moves from the crop to a weed, the herbicide will no longer be effective. If is extremely expensive to develop a new herbicide, and it is not in the best interest of the company to have it become ineffective in a short time. This economic incentive for selective use of HTCs in areas without cross-compatible weed species will be important. In addition, farmers will now be interested in products that may create worse problems than they will solve.

I have tried to share with you my overview of the development and use of HTCs. As you can readily discern, my perspective is strongly influenced by living most of my life in the Midwest where herbicides are widely used. The perspective may have been different if I was concerned about forestry, vegetable crops, or other plant species or geographical areas. Other meeting participants will have different perspective to share in the meeting workshops.

PART II

HERBICIDE TOLERANT CROPS: A BENEFIT-RISK ASSESSMENT

The Biotechnology Program at Iowa State University deals not only with the scientific and technical aspects of the application of biotechnology, but also with its possible social and economic impacts. Devoted to the probable risks and benefits of introducing herbicide tolerant crops (HTCs) in Iowa, this paper is the result of a conference held at Iowa State University in late October 1990, which included participants from academia, federal and state government, industry and other selected organizations. The final content of this report is the responsibility of the Office of Biotechnology, Iowa State University, and reflects the opinion of the majority of the participants. Participants at the conference did not necessarily concur with all aspects of the report.

INTRODUCTION

On October 29–31, 1990, a conference was held at Iowa State University to conduct a benefit-risk assessment of the introduction of herbicide tolerant crops in Iowa, and a document prepared to communicate the results of the meeting to the state's policymakers and general public.

Herbicide tolerant crops (HTCs) are crops that tolerate a certain amount of a selected herbicide without damage. Any crop variety must be tolerant in order to avoid injury when a herbicide is applied. Therefore, current crop varieties are herbicide tolerant if they are grown successfully in areas to which a herbicide is applied. Currently, between 30 and 40 herbicides are available in Iowa, but lack of tolerance in some crops restricts herbicide use in the typical soybean/corn rotation.

Increased herbicide tolerance in crop varieties can be developed through various techniques of biotechnology, including tissue culture and genetic engineering. In the context of this report, HTC assumes the use of biotechnological techniques to increase the level of tolerance of a crop to a herbicide beyond what is currently present. Successfully developed HTCs would allow certain herbicides that currently damage corn or soybeans to be used for weed control in those crops.

The purpose of the conference was *not* to determine the advisability of continuing the use of herbicides in agricultural production in Iowa. Instead, the focus was: If herbicides continue to be used in Iowa, what is the most probable role of HTCs? Conducting an assessment of the benefits and risks of HTCs assisted the participants in defining the new technology's probable role. The charge given to the conference participants was to concentrate on three areas: 1 - population ecology and genetics; 2 - environmental quality and consumer health; 3 - socioeconomic impacts. During this process, participants were asked to determine the issues on which they could reach a consensus of agreement, to clarify the issues on which they disagreed and why, and to identify areas in which more research is needed.

A brief overview of the major areas of consensus:

—Weed control is critical for crop productivity. Development of effective weed control technologies will continue to be essential for Iowa agriculture.

—Some weeds cannot be controlled effectively using current herbicides because killing the weed also damages the crop plant. Introducing a gene for herbicide tolerance into the crop plant would allow the use of a herbicide that kills the weed without damaging the plant.

—Some herbicides are less desirable for the environment than others. Herbicide tolerance should be pursued only for those herbicides that have minimal negative impact on the environment. Iowa should not seek to increase or even maintain at current use levels those herbicides that are unfavorable to the environment or human health by developing crop tolerance to them.

—The amount of herbicide applied per acre could increase or decrease with the use of HTCs, depending on which herbicides crops are genetically modified to tolerate and how HTCs are incorporated into weed control practices. Since 97 percent of corn and soybean acres in Iowa are currently treated with herbicides, it is logical to assume that the number of corn and soybean acres to which herbicide are applied could only increase by three percent. HTCs will likely lead to displacement of current herbicides with others that potentially may be more favorable for the environment.

—All major crops are already tolerant to many herbicides. Development of HTCs is unlikely to dramatically change the amount of herbicide use for most major crops, but will expand the types of herbicides available for weed control.

—Many farmers use pre-emergence herbicides as "insurance" against potential weed problems. With HTCs, herbicides could be used only when necessary in post-emergence applications, thus reducing the total amount applied.
—Herbicide tolerant varieties of crops should be evaluated by federal regulatory agencies to ensure the safety of the food supply. Current regulations of the EPA require that herbicide residues and their health risks be determined as part of the registration process for any new herbicide or HTC.
—Herbicide tolerance in plants should not be engineered to the point that farmers can over-apply herbicides without harming the crop plants. It is advisable to develop crops that are tolerant to only the minimum amount of herbicide necessary to control weeds, with crop damage occurring if excessive amounts are used.
—Genes for herbicide tolerance should not be introduced into crops where there is a wild weed species with which the crop could intercross. In Iowa, there is no evidence that the movement of a herbicide tolerance gene from corn or soybeans to weeds would be a problem because there are no known weed species in the state with which corn and soybeans can intercross. Each crop into which a herbicide tolerance gene is introduced must be assessed separately for the likelihood that intercrossing with weed species will occur.
—Herbicides used with HTCs should be rotated. If the same or similar herbicides with the same mode of action are used year after year on the same ground, the weeds that withstand it the best will be the ones that survive and produce seed. This natural selection will occur every season. Eventually, these weeds will be numerous enough to again cause a problem.
—It is the responsibility of Iowa State University and its Extension Service to continue to educate farmers about the alternatives for effective, long-term weed control. With so many new technologies becoming available to farmers, only one of which is herbicide tolerant crops, farmers must learn how to integrate the various options into the best management plan for their farm. Optimum weed management strategies should rely on an integrated approach, including crop rotation, cultivation and the minimum use of herbicides.

—Iowa State University should not be involved in the actual development of HTCs. Iowa State University and other public research and educational institutions should continue to invest in researching the appropriate use of herbicides in an overall weed control strategy.

—More research is needed in several areas, including why and how herbicide tolerance evolves or does not evolve in a plant; how to investigate the real dangers associated with herbicide HTCs; how genes flow from crops to weeds; any long-term health risks; possible unintended changes in plant nutrients, natural toxins, or allergens when plants are genetically engineered and treated with herbicides; and integrated weed management systems for agriculture that are unprofitable for the private sector to research or develop.

PURPOSE OF THIS DOCUMENT

The purpose of this document is to communicate to Iowa's policymakers and the public the results of the Iowa State University meeting. It is not written in the highly technical style of a scientific journal because it is not, primarily, for scientists. Neither is it written in the form of a recommendation because its purpose is not to advocate one position over another. Instead, this document is written to inform Iowa's policymakers and public of what some of the top experts believe are the benefits and risks of using herbicide tolerant crops in Iowa.

This document poses questions asked by those examining the role of herbicide tolerant crops in agricultural production. It discusses the answers to those questions as they evolved at the Iowa State University meeting. It pinpoints the questions that remain unanswered. It identifies areas of agreement and disagreement. It is, in short, a written record of the collective expertise of those who attended the meeting.

ISSUES AND ANSWERS

1. *What could HTCs do that current herbicide use on crops cannot?*

Some weeds cannot be controlled efficiently using current herbicides because killing the weed also damages the crop plant. Introducing a gene for herbicide tolerance into the crop plant would allow the use of a selective herbicide that kills the weed without damaging the crop. This will expand the types of herbicides available for weed control in a crop.

2. Will HTCs greatly increase herbicide use in Iowa?

The most recent study available, a cooperative effort between the Iowa Crop and Livestock Reporting Service (now Iowa Agricultural Statistics) and Iowa State University's Cooperative Extension Service, estimates that 97 percent of Iowa corn and soybean acreage receives at least one application of herbicide each growing season (Wintersteen and Hartzler, 1987). There was little change in this percentage during the period of the study from 1979–1985. Since only about three percent of corn and soybean acres do not have herbicides applied now, the number of acres to which herbicides are applied could increase only slightly.

The majority opinion of meeting participants was that the amount of herbicide applied per acre in certain situations could decrease with the use of HTCs. Many farmers use pre-emergence herbicides as "insurance" against weed development later in the growing season. This amounts to applying herbicide for a weed problem before knowing for certain if the weed problem will develop. Environmental factors and cultivation practices affect whether, and how severe, a weed infestation will be. HTCs can allow better post-emergence weed control, allowing the farmer to "wait and see" if a weed problem develops before opting for chemical control. With HTCs, post-emergence herbicides could be used as a clean-up rather than as a preventative.

Scouting a field for insect and weed problems is already being used as an alternative to blanketing a field with herbicide or insecticides as "insurance." Commercial businesses offer the service or knowledgeable farmers can scout their own fields.

All major crops are already tolerant to many herbicides. Development of HTCs is not likely to change the amount of herbicide use, but will expand the types of herbicides available for weed control in a crop. HTCs offer the potential to make more environmentally sound choices among herbicides in certain situations.

3. Must HTCs have herbicides applied to them in order to thrive?

One misperception is that a HTC somehow draws its nutrition from herbicides and must be "fed" herbicide in order to live. This is not true.

People who receive a flu shot in the autumn to protect themselves against influenza do not need to have the flu in order to live. However, they can usually withstand exposure to the influenza virus without catching the disease.

It is a similar situation with HTCs. A HTC can withstand an application of herbicide that kills weeds in the field; it does not need the herbicide to live.

4. Will the use of HTCs affect the safety of the food supply?

The conclusion of conference participants was that herbicide tolerant varieties of crops should be evaluated on a case-by-case basis to guard against any potential problems. The International Food Biotechnology Council (IFBC), an industry organization, has proposed a set of guidelines for use by food regulatory agencies. The guidelines are titled *Biotechnologies and Food: Assuring the Safety of Foods Produced by Genetic Modification* (IFBC, 1990a).

The majority opinion was that these guidelines should be applied to HTCs just as they would be to any new food product developed through biotechnology. The IFBC published in *Regulatory Toxicology and Pharmacology* a summary of major issues regarding safety assurance of foods produced by the use of biotechnologies (IFBC, 1990b).

Approval of EPA must be obtained before a herbicide is registered for use on a crop. Environmental Protection Agency registration requirements should also apply to the review of HTCs for commercial use. Participants agreed that two types of tests that should be conducted on new HTC varieties are feeding trials and an evaluation for toxicants. Feeding trials are carefully controlled experiments in which animals are fed grain produced by HTCs and are evaluated for any ill effects. Evaluation for toxicants means the grain would be examined for harmful substances.

The participants agreed that herbicide tolerance in plants should not be engineered to the point that farmers can over-apply herbicides without harming the crop plants. The ideal is to have a crop plant that can withstand several times the normal herbicide strength to insure tolerance, but not 100 times the normal strength. At lower levels, plants metabolize or break down the herbicide, and it is not likely that any would survive in the plant to be passed into the food chain. If a plant could tolerate an extremely high level of herbicide, it is possible that the plant would not be able to break it all down; some might accumulate in the plant and be passed into the grain. The participants agreed it would be advisable to develop crops that are tolerant to only the minimum amount of herbicide necessary to control weeds, with crop damage occurring if excessive amounts are used.

Developing HTCs that suffer injury when herbicides are over-applied also avoids the problem of farmers increasing the herbicide application rate because they do not have to worry about carryover herbicide residues in the soil if the following crop is also herbicide tolerant.

5. Can the gene for herbicide tolerance accidentally move from the crop into a weed species, making the weed tolerant to the very herbicide used to control it?

For genes to move from one species to another through natural crossing, the two plant species must be closely related. In Iowa, there is no evidence that this would be a problem for corn and soybeans since there are no known weed species in the state with which corn and soybeans can intercross.

However, the majority opinion of the group was that genes for herbicide tolerance should not be used in crops where there is a wild weed species with which the crop could intercross. For example, shattercane is a weed in fields of its relative, sorghum. Although it would be of short-term benefit if a herbicide tolerance gene could be inserted into sorghum to allow it go be unharmed by a herbicide that kills shattercane, there could be long-term problems. The gene for herbicide tolerance could move from the sorghum into its shattercane relative, making shattercane tolerant to the herbicide.

Oats and horticultural crops in Iowa could encounter similar problems if herbicide tolerant genes were developed for them since they have wild relatives in the state that growers classify as weeds.

The group agreed that each crop into which a herbicide tolerant gene is introduced must be assessed separately for the likelihood that intercrossing with weed species will occur.

6. Will HTCs promote the development of herbicide tolerant weeds?

In addition to the accidental movement of a herbicide tolerance gene from a crop into a weed related to it, there is the possibility that if the same or similar herbicides with the same mode of action are used year after year on the same ground, the weeds that withstand it the best will be the ones that survive and produce seed. This natural selection will occur every season, and eventually these weeds will be numerous enough to again cause a problem. The only way to prevent this from happening is to utilize an integrated weed control strategy, including rotation of herbicides with different modes of action, crop rotation and cultivation. This principle also applies to the appropriate use of currently available herbicides.

The participants' majority opinion was that it is the responsibility of the chemical industry and institutions such as Iowa State University and its Extension Service to educate farmers about the rotation of herbicides with different modes of action as part of an integrated strategy for weed control.

7. Will HTCs reduce the genetic diversity of weeds that are wild relatives of HTCs but are not a problem in the crop field?

Genetic diversity within a plant species allows adaptation to and survival of changing conditions. For example, a potato plant that expresses a gene for drought tolerance is more likely to survive a dry spell than the potato plant next to it that does not express that gene so strongly. Nature keeps many different genes in a plant species' repertoire. If a particular gene is needed for the species' survival at some point in time, it will be there.

If wild relatives of herbicide tolerant crop plants somehow receive the herbicide tolerance gene, will there genetic diversity be affected? No. For example, a wild relative of corn is *Tripsacum*. In Iowa, *Tripsacum* is not a weed that appears in corn fields so it is not exposed to herbicide. The majority opinion of the group was that if *Tripsacum* were to somehow receive a herbicide tolerance gene from corn, there would be no problem. Since *Tripsacum* is not a field weed and would not be exposed to herbicides, there should be no selective advantage for the *Tripsacum* plants that have the herbicide tolerance gene over those plants that do not. Both *Tripsacum* with and without genes for herbicide tolerance should survive and their genetic diversity unaffected.

It was the majority opinion that genes for herbicide tolerance quite possibly could be passed to weeds. However, the ecological consequences would be minimal because the weed with the introduced gene will not have a higher survival rate than other weeds in the wild, only in the field. Because natural selection is unlikely to favor the spread of herbicide tolerance genes and alter the genetic structure of wild populations of weeds, it is unlikely that existing genetic diversity would be completely lost from weed species.

8. Would the development of herbicide tolerant trees increase the use of herbicides in forestry?

Forestry is a unique crop situation because conventional herbicide and crop rotation practices usually are not utilized. The same trees stay in the same soil for years. Herbicides are currently only used on a fraction of forest acres, but herbicide tolerant (HT) trees are being developed to permit more extensive herbicide use. Some participants believed that HT trees would have little environmental impact since herbicide use on trees would be limited to the first year or two after planting. Other participants felt that expanded herbicide use in forestry was environmentally unacceptable because forests are commonly used for conservation and recreation, as well as for harvesting timber.

Other participants believed there could be a problem with HT perennial plants if multiple applications of the same herbicide were used over a short period of time in short rotation forestry.

One suggestion was that HTCs in trees should be coupled with a sterility system so the herbicide tolerance gene does not spread into the native population of trees.

9. How would HTCs affect the way farmers manage their crops?

For farmers who choose to use herbicides as a weed control method, HTCs could increase their flexibility by increasing the herbicide choices and timing of applications.

The participants observed that public sector research must put more emphasis on integrated weed control strategies that minimize herbicide use. With so many new technologies becoming available to farmers, only one of which is herbicide tolerant crops, farmers must learn how to integrate the various options into the best management plan for their farm. Optimum weed management strategies should rely on an integrated approach, which may include crop rotation, cultivation and the minimum use of chemicals.

10. Who bears responsibility for educating farmers about HTCs and their appropriate use?

Attendees saw a need for Extension Service personnel to increase their own knowledge of HTCs and the appropriate use of HTCs. The knowledge could then be passed to farmers. It was agreed that routine use of herbicides should not be promoted. Instead, farmers should understand that herbicides should be used only when necessary and at the rate needed.

Others noted that chemical companies have a great deal to lose in future profits if resistant weeds evolve from the misuse of herbicides, and that there is a need for companies to educate their salespeople and customers on the importance of proper herbicide use.

11. Can farmers with small operations and farmers with large operations both access HTC technology equally?

The majority of participants agreed that the cost of HTCs would not prevent those with small farm operations from using them. However, HTCs might promote larger farms since a farmer might be able to handle more acres with fewer people. Yet other factors such as increased mechanization, the exodus of farm youth from the family farming business, and the greater economic stability of off-farm work may also contribute to larger farms in Iowa.

12. Could HTCs increase yield and contribute to a price decline that could force small farms out of business?

The majority opinion was that HTCs would increase yields only where it became possible to control weeds that are not being controlled now. One member of the group estimated that this might involve 5 percent of the crop acres in Iowa. The only reasons for farmers to use HTCs would be for improved weed control, to lower the cost of production, to reduce the amount of pre-emergence herbicides used, or to provide greater flexibility in the use of more environmentally favorable herbicides.

Wide adoption of HTCs in Iowa might even result in a "yield penalty" for farmers who choose to use them. For example, suppose a company begins research to insert a herbicide tolerance gene into its highest yielding corn variety in 1991. With current technology, the company will spend five to seven years incorporating the gene. By the time the variety is ready for release in 1996, other higher yielding varieties without the herbicide tolerance gene could have been developed. The farmers who choose the 1991 herbicide tolerant corn must pass up the 1996 higher yielding variety and penalize themselves as far as yield is concerned.

13. What are other possible socio-economic impacts?

The possible socio-economic impacts of the introduction of HTCs for Iowa include the following:

—No large impact on agriculture is expected.
—Changes in grain quality are unlikely for corn and soybeans.
—There may be a potential small shift to expansion of farm size due to freeing up labor, but other factors are likely to affect changes in farm size more than HTCs.
—There will probably be a neutral impact on crop yields, unless weeds can be controlled more effectively.
—Farmers are unlikely to be willing to pay higher seed costs, unless there is a savings in herbicide costs or better weed control is achieved.
—Increased options and flexibility in herbicide use are likely.
—HTCs may foster improved soil conservation if they provide more effective weed control in conservation tillage systems.
—An increased level of management skills will be required of farmers to coordinate crop variety and herbicide selection and use.
—The competitive structure of the seed/chemical industry could change.
—HTCs could lead to new regulatory constraints and costs for industries and farmers.

14. Should public research funds be used to develop HTCs?

The majority opinion was that Iowa State University should not be involved in the actual development of HTCs. Iowa State University and other public research and educational institutions should continue to invest in researching the appropriate use of herbicides in an integrated strategy for weed control.

The socio-economic experts suggested that public sector research may shift to developing weed control technologies other than HTCs since industries are doing the HTC research.

15. Will HTCs allow increasing amounts of undesirable herbicides to enter the ground water, soil, and air?

The meeting participants agreed that herbicide tolerance should be pursued only for those herbicides that have minimal negative impact on the environment. It was their belief that Iowa should not seek to increase—or even maintain at current use levels—those herbicides that are unfavorable to the environment or human health. HTCs for more environmentally favorable herbicides could replace the more persistent herbicides that can become environmental pollutants.

Since the increased use of one type of herbicide might cause environmental contamination and impact wildlife, it is important that herbicides be used in a rotation so chemicals with the same mode of action are not applied year after year to the same field.

Attributes suggested for herbicides used in conjunction with HTCs include:

—Low toxicity to non-target species, including humans and wildlife.
—Low residues in the environment.
—Non-toxic residues in the crop and food.
—Low threat to ground water, surface water and air.
—Low use rates.
—Appropriate degradation (breakdown) with benign breakdown products.
—Cost effectiveness.
—Compatible with alternative weed management strategies.
—Compatible with technology improvements in the way the herbicide is applied.
—Increased reliability of weed control accompanied by improved crop yields.

16. What are the gaps in knowledge about HTCs; what is an appropriate research agenda?

More research is needed on why and how herbicide tolerance evolves or does not evolve in a crop or weed. The majority opinion was that the amount of basic weed science research and the dollars to support it should both increase.

Perceived versus Real Risks—Society must sort out the perceived dangers versus any real dangers associated with HTCs. For instance, many members of the public perceive that HTCs will cause increased herbicide use in Iowa. Is this a legitimate concern, or is present herbicide use so high that HTCs are unlikely to have a significant impact?

Many of the participants at the meeting viewed the perceived risks of HTCs as the same risks faced today in agriculture when a new herbicide is introduced, such as potential use of higher herbicide rates by farmers, increased dependency on single chemical exposure, new or increased toxic chemical residues in crops, altered plant properties, food residues and less diverse farming systems. In assessing the benefits and risks of HTCs, a distinction must be made between risks more or less unique to HTCs and risks applicable to any herbicide use.

Gene Movement—More information is needed regarding if and how genes flow from crops to weeds. Is it possible for genes to transfer in ways other than by natural crossing? Is there a non-sexual method of gene transfer? A study should be done to monitor plant population and gene shifts before and following the introduction of an HTC.

Should regulatory agencies try to prevent HTC use in areas where weeds are related to the HTC? Should the technology, once developed, be kept from other countries where it should not be used due to the presence of weed relatives?

Environmental and Health Considerations—Most health risk concerns for HTCs will probably by similar to those associated with currently available herbicides. How will certain HTCs affect the environment? Will there be an increased risk to farm workers who work with HTCs? Are there synergistic factors that work together to increase herbicide health risks? For example, does a person's diet interact with herbicide exposure to increase the risk for disease more than would either diet or exposure alone?

Unintended Plant Responses—Research needs to be done to ensure that plants will not develop unintended changes in nutrients, natural toxins, allergens, etc., when they are genetically engineered and treated with herbicides.

Farm Management Systems—While research on chemical weed management through the use of herbicides is important, some participants at the meeting believe research support in the universities has been skewed in this direction in the past. They suggested that Iowa State University and other agricultural institutions should develop research agendas stressing viable, overall, integrated management systems for agriculture, especially those systems that are unprofitable for the private sector to research.

Can herbicide tolerance genes be developed that encourage rotation of herbicides? Should Iowa avoid introducing tolerance to a herbicide in both corn and soybeans?

Socioeconomic Questions—The sociologists and economists present at the meeting defined the following research questions:

—Which is the best allocation of Iowa's resources, herbicide tolerance or the development of resources?

—Who decides the research priorities for industries and universities?
—What effects will HTCs have on related research and product development?
—How can the self-interest factors of land-grant colleges, the chemical industry and the seed companies be evaluated?
—What is the appropriate method to use in evaluating impacts of a new technology like HTCs?
—What are the public's values and beliefs regarding HTCs?
—What factors influence farmers' and the public's acceptance of HTCs?
—What are the educational needs of farmers and the public concerning HTCs?

REFERENCES:

IFBC, 1990a. International Food Summary. *Biotechnologies and Food: Assuring the Safety of Foods Produced by Genetic Modification.* Printed as a supplement to *Regulatory Toxicology and Pharmacology,* Volume 12.

IFBC, 1990b. International Food Summary. *Biotechnologies and Food: Assuring the Safety of Foods Produced by Genetic Modification.* Printed as a supplement to *Regulatory Toxicology and Pharmacology,* Volume 12.

Wintersteen, Wendy and Robert Hartzler. 1987. *Pesticides used in Iowa crop production in 1985.* Cooperative Extension Service, Iowa State University of Science and Technology and the United States Department of Agriculture cooperating. Pm–1288.

Part II reprinted from *Iowa State University Biotechnology Update.* 1991. A Benefit/Risk Assessment for the Introduction of Herbicide Tolerant Crops in Iowa, Executive Summary. 5 (Feb. 1991):1-14.

BIOLOGICAL CONTROL
MAKING IT WORK

BRIAN F. CHABOT

Director of Research
Cornell Cooperative Extension
Agricultural Experiment Station
Cornell University
Ithaca, NY 14853

PART 1: SUMMARY & RECOMMENDATIONS

On April 4 and 5, 1991, 52 individuals from academia, the private sector and state and federal agencies met to discuss future directions for biological control of pests in the United States. The objective of this meeting was to identify the major steps that need to be taken to expand the availability and use of biological pest controls. The group agreed that the range of available pest control tactics needs to be expanded and that biological controls, broadly defined, need to be more aggressively developed. What emerged from the workshop was a rather lengthy list of actions. While these were eventually combined into four major areas, it seems clear that multiple steps must occur involving a number of organizations in order for biological control to become more effective. There was no single step that alone could produce a major advance, however, there was a high level of certainty that additional investments in addressing the needs would lead to concrete payoffs of considerable advantage to the nation.

The workshop involved a select group of invited participants representing major interested parties. Since we relied heavily on informal discussions among the entire group and subgroups, the total number of participants was intentionally limited. Four background papers were distributed prior to the workshop, and they follow in this background report.

The workshop consisted of three periods of general discussion among all participants and two breakout sessions where subgroups developed ideas in the areas of technology development, commercialization, regulation and adoption. In the first breakout session, groups identified and prioritized issues that were limiting fuller utilization of biocontrol practices. The second breakout session identified actions that could resolve the limitations. While the subgroups focused on different areas, what emerged were several common areas of concern. Stated as actions, these were:

1— Develop a national agenda, leadership and policy structure including:
- National and state policies encouraging biocontrol practices as part of total pest management.
- Mechanisms to coordinate activities
- Leadership identified in each state
- Regulatory policies different from those for chemical pesticides

2— Aggressively develop bio–based pest control
- Understand fundamental biology of the pest, biocontrol agent and ecological system
- Develop production, storage, delivery and application methods
- Address scientist base, funding structure and other incentives to accelerate programs
- Encourage farmer and extension participation
- Develop and integrated, systems approach in commodity–based programs

3— Expand communication and education efforts
- Improve understanding by scientists of current regulatory procedures
- Encourage more constructive discussions among interested and involved parties
- Respond to specific information needs—fact sheets, demonstrations
- Develop improved extension and consultant training
- Increase public/consumer education about pest–management practices

4— Provide incentives to facilitate adoption
- Need appropriate economic and environmental assessment of different practices
- Address or change cosmetic standards
- Establish clear regulatory environment
- Reduce cost of regulatory compliance for niche products
- Develop universal protocols for risk assessment
- Establish protocols for property rights protection for biocontrol practices

SUBGROUP REPORTS

Four subgroups, in the areas of technology development, commercialization, regulation and adoption, each developed a list of their concerns and potential solutions, which are presented in the following text.

TECHNOLOGY DEVELOPMENT

Generate, expand and improve the appropriate use of genetic resources while maintaining diversity

Plants and soils support a great diversity of organisms with the potential to provide defense of the plants through natural biological control. Different management systems and agroecosystems support unique mixtures of organisms with different potentials to provide biological control. In the past, most attention has focused on genetic diversity, expansion, improvement and appropriate use of plant genetic resources. This effort should continue, but significantly more attention should be given to the genetic resources represented by beneficial organisms and populations of organisms. Associated with the crop, including beneficial plant and soil associated microorganisms, natural enemies of insect pests, nematodes and plant pathogens, or insects and pathogens with potential to attack weeds. To this end:

a. Establish and maintain germplasm of biocontrol agents
b. Genetically assess potential agents
c. Genetically improve biocontrol agents
d. Isolate resistance genes and function of genes from plants and pathogens
e. Screen potential new biocontrol agents

Understand mechanisms of biocontrol at the molecular, genetic, organismal population and ecosystem levels

Basic research is needed on mechanisms of biological control at the molecular, genetic, organismal, population and ecosystem levels as a means to understand and thereby make greater use of biological control. The following phenomena need additional research:

a. community structures
b. timing and phenology of pest and host
c. epidemiology

d. key traits affecting interorganismal interactions and management

e. gene function(s)

This research could result in changes in cultural practices to favor more natural biological control or a specific introduction (one–time, occasional or repeated as necessary) of a biological control agent or mixture of agents to supplement, expand or create a biological control system. The exact deployment would be determined by the nature and genetic control of the mechanisms deployed. It could include a mechanism(s) deployed in a organism or population or organisms associated with the crop, or when fully understood at the molecular and genetic level, it could be deployed as a transgenic plant.

Develop implementation technologies

New technology needs to be developed in support of biocontrol implementation for all categories of biocontrol organisms. Current inability to culture and economically mass produce biocontrol organisms limits the development and use of these organisms to control pests. In addition, inconsistency in quality, efficacy and reliability of these products affects producers' acceptance of biocontrol technology. Formulation and storage technology are critical issues and must be improved for the effective distribution and delivery of biocontrol agents or bio–based products. Subsequent to production, storage distribution and delivery comes the introduction and management phase where efforts focusing on the manipulation and management of natural biological organisms and their life cycles must be improved. New ecologically and genetically based technologies must also be integrated effectively with naturally occurring and introduced biocontrol organisms as well as other pest management tactics. Specific needs include: Formulation of agent, mass production, storage systems, quality control, delivery methods and equipment, introduced organism/gene products and the management of natural ecological cycles.

Assess and evaluate technology

Biocontrol agents and strategies must be evaluated for their performance and reliability as well as any consequences of their use across a range of agroecosystems. Consistent and reproducible technologies should be encouraged. The consequences of using biocontrol technology must be evaluated, particularly the technology's impact on non–target organisms as well as the rate and spread of the agent. Additionally, how a biocontrol technology fits into crop protection practices should be evaluated. For example an analysis of the cost and efficiency of this technology relative to conventional chemical or plant breeding alternatives is necessary. The various producer/consumer groups which will experience benefit or harm need evaluation with respect to: biological and ecological impacts, risk and safety with respect to environment, socioeconomic aspects and general assessments of biocontrol technology.

Complement existing pest control systems and develop long–term alternative systems

In the short term, biocontrol must be integrated into existing pest and disease management technologies, including chemical controls. Biocontrols are also needed in both the short and long term to limit the effects of those pests and diseases for which currently there are no alternatives. Examples include certain virus diseases of plants, especially insect–vectured viruses and soil-inhabiting pests such as nematodes, soil–borne pathogens and soil insects that now are only controlled by uneconomical crop rotations. Gradually, however, within the next 25–50 years, biocontrol technologies must be considered as the main approach to crop protection, including cultural practices to manage natural beneficial cycles of biocontrol.

COMMERCIALIZATION

Put biocontrol on the national agenda

Biocontrol approaches to pest control in plant production, animal production and public health in the United States have not been a high priority issue. Several compelling reasons at this time to give biological control a high priority:

—Pest control procedures with less environmental impact in areas such as toxic chemical residue in groundwater and soils.

—Pest control agents to substitute for the increasing number of insecticides, fungicides and even some herbicides to which pests have developed resistance

—Bio–based pest control agents and procedures to reduce the consumer concern (perceptions) about the risk of chemical pesticides in food

—Biological approaches to discovery needed pest control agents since there is a decreasing rate of discovery of chemical pesticides

—Need to satisfy largely unmet needs for pest control problems such as soilborne pests and sucking insects

—Improve international competitiveness with improved productivity of commodity agriculture crops and new international markets for agribusiness through biological control agents.

—Improve safety of pest control manufacturing workers and pest control applications by replacing synthetic chemicals with biologicals

—New biological technology to facilitate the research and development needs for biological control agents

Actions to put biocontrol in the national agenda:

—Communicate the need for pest control and the potential of biocontrol to national decision–makers

—Forge a coalition of stakeholders for communicating and establish ing working relationships

—Implement priority items

Address the paucity of proven biocontrol candidates and fundamental knowledge

To date, only a limited number of candidate organisms have been identified as potential biocontrol agents, a charge exemplified by the fact that less than one percent of the worldwide pesticide market is represented by biocontrols.

Additional candidate biocontrol agents must be identified if biological control is to take its potential and proper place as a means of controlling pests and pathogens. An extensive program must be implemented to expand the list and number of potential biocontrol agents and as potential biocontrol agents are identified, they must be characterized as to mode of pest action and the genetic basis of that action. This will result in the basic knowledge required to make genetic improvements of biocontrol organisms. Further, to expand upon the genetic sources of resistance, resistance genes should be identified, isolated and characterized from a variety of sources, including plants and pathogens.

The fundamental knowledge of the action mechanisms of biocontrol agents' at the molecular, genetic, organismal population and ecosystem levels will permit implementation of integrated systems for maximum performance of biocontrol agents. Based on studying community structure, epidemiology, traits affecting interorganismal interactions and management, gene functions and the life cycles of both pest and biocontrol agent, this basic understanding will permit proper timing of biocontrol agent applications and will result in an integrated systems approach to plant and animal protection and the control of pests impacting public health.

To accomplish these goals, a pool of highly trained technical and business personnel must be available to develop the base technologies, provide the biocontrol products and effectively bring them to market. To initiate this process, all existing biocontrol educational information should be pooled to provide a common informational source to all interested parties.

Insufficient formulation and delivery technology for biological control systems is of concern

The lack of adequate research efforts directed toward formulation and delivery technologies is a major limitation to the implementation of commercial biocontrol products. Although adequate research funding can often be identified for supporting both basic and applied biocontrol research, funding is generally not available for mid–level research. The fundamental differences between chemical pesticide formulation, production and delivery systems and those of biocontrol agents is an additional difficulty.

Research production technology

Research and development for microbes is needed to develop appropriate fermentation technology for "unusual" bacterial and fungal species not typically used by the fermentation industry. Additionally, the focus has to be on production measured by total bio–activity, not simply yield. Research and development for viral agents needs to focus on cost–reduction and generic technologies applicable to many viruses. Research and development on macro–agents needs to focus on the low–cost production and maintenance of viable organisms.

Research formulation technology

Unlike the formulation of chemical pesticides, biological agents need to be formulated in such a way as to maintain visible populations of the agent in a physiologically active state. Ideally, formulations should also be compatible with existing application equipment and consistent with the ecology of the agent. Special efforts should be directed toward the development of formulations that will extend the shelf life and storage capabilities of biological agents.

To accomplish the above, appropriate funding needs to be identified in order to train quality individuals in such mid–level technologies. For example, the fostering of industry/academic relationships may lead to the exchange of personnel.

Address the lack of clear, consistent regulations for field testing and commercial registration

A number of regulatory issues impact the field testing, registration and commercialization of bio–based pest control agents. Most importantly, the regulations (risk assessment for field testing and registration) should be based on the end product rather than the process by which the product (pest control agent) was produced. Because an organism is genetically engineered or genetically modified in some way does not necessarily mean it is a higher risk than a naturally–occurring organism.

Approval for field testing should be based on familiarity, containment and risk, as stated in the National Research Council's 1989 report. This would expedite field testing for biocontrol agents, since large data packages may now be required for resting familiar organisms that can easily be contained.

Another issue seriously impacting development of bio–based pest control agents is the inconsistency between state and federal regulations. Harmonizing these regulations would expedite commercialization. At present, regulators have few incentives to adjust the requirements for bio–based pest control agents based on risk, and these incentives need to be identified in order to streamline the regulatory process.

Address the lack of trust for science and technology–based solutions in pest management

Science and technology have greatly advanced our ability to manage pests of human health, food and fiber at low cost and with great safety. Nonetheless, as in all human endeavors, some efforts (such as DDT) have been less successful than others. Unfortunately, public attention has often focused on these missteps rather than successes, leading to a lack of trust for novel technologies in pest management. Particularly with respect to biological pesticides and transgenic plants, the technologies are very abstract and poorly understood by the public, leading to apprehension about the possible hazards of the technology. All too often, attention focuses on the genetic process rather than the product. In many cases, the criticisms raised about novel pest management tools betray a fundamental misunderstanding of the procedures used in biotechnology. Because no technology can be completely risk–free, it is first essential to educate and communicate with the public about the risks and benefits not only of biotechnological pest controls, but also to compare these with the risks of the technologies that would be replaced. Second, because at least some of this information is highly technical and not all of the interested public will have the time to digest it, there is a need for groups without vested interests but strong technical expertise to publicly give their own appraisals of the technology. Only after these steps have been taken can the public be given an adequate opportunity to make their wishes known

Protect intellectual property

The United States' patenting system is well established for chemically based control systems and therefore provides incentives for investment in research and development by giving the investor a true–limited monopoly on the acquired intellectual property. Similar protection is not well developed for biocontrol agents; the patent office has either been slow processing applications in granting rights, or is not able to grant exclusive rights (macroorganisms, some microorganisms).

Proposed actions:

a. Expedite review of pending patents for inventions deemed patentable.
b. Establish a protective system for biocontrol agents not deemed patentable loosely modeled after the Plant Variety Protection Act. Contrary of PVPA, this act would not contain farmers' rights to propagate.
c. Establish commercial incentives for research, development and commercialization for biocontrol agents with minor use or niche markets modelled after the "Orphan Drug Act."

REGULATION

Support agricultural commodities

Agricultural commodity support policy has to include environmental consideration. Several factors currently create disincentives for adopting more flexible cropping systems that integrate biological control and thus should be eliminated.

The current structure of farm commodity support programs penalizes the adoption of flexible cropping systems that are more conducive to the adoption of biologically–based control of crop pests.

Payments producers receive under these programs is calculated by multiplying deficiency payment *times* program yield *times* base acres. Both program yields and base acres are administratively set as running five year averages of actual yield and acres planted to program crops. There is a strong incentive for producers to 1–maximize yields of program crops (which may discourage the adoption of biologically based pest control strategies that may result in lower yields even though these practices may be more cost effective) and 2–maximize acres planted to program crops (which discourages the adoption of rotations, strip cropping, intercropping or other cropping systems that are more conducive to biocontrol).

Current commodity policy should be reformed to eliminate these penalties to adopters of more biologically–based pest control practices.

We recommend that NABC develop a brief position document that explains the way current commodity programs penalize the adoption of cropping systems more conducive to biocontrol. This position document should be submitted to USDA, EPA, OMB and the public to encourage reform of current commodity program regulations.

Increase communication

The communication between different organizations involved in making biological control a viable alternative have to improve.

Communication about the regulatory process and requirements for approval of biocontrols needs to be improved. To improve communication between federal agencies and the scientific community, NABC should identify the information needs of the scientific community, develop materials (such as fact sheets or other primers) and prepare a plan for dissemination of these materials. EPA should develop and disseminate guidance or models for successful data waiver rationales. NABC should investigate opportunities for developing "regulatory" ombudsmen who could serve as sources of expert advice on the regulatory process to the scientific community. In doing this, NABC should consider whether and how such ombudsman positions could be developed within the academic community, scientific and other professional associations, or within government agencies.

Federal and state regulatory agencies need to examine whether their current regulatory processes work effectively together to provide adequate safety evaluations of biocontrols without unduly impeding the research, development and commercialization of these products. The scientific community, industry, public interest groups and regulatory agencies need to examine the existing opportunities and mechanisms for informing the public about biocontrols and providing adequate and effective opportunities for public input in the regulatory decision making process.

In order to achieve these goals:

a. Identify the regulatory information needs of researchers

b. An answer to these questions should be presented, preferably through "easy to use" information; e.g., fact sheets that summarize the regulatory procedures. This could be done by the NABC.

c. Institutions (EPA, USDA, ...) provide regulatory models for application procedures.

d. A guide dissemination of their information should be provided.

e. An "ombudsman" could be installed in an appropriate organization to help process the questions, information and applications.

f. The NABC should develop a (short) rationale that explains this point and this should be distributed to the following agencies:

Congress
Environmental Protection Agency, U.S. Department of Agriculture
Office of Management and Budget
farm organizations
public interest groups

Risk assessment protocol is very important

There is a need for the development of a universal and coherent risk assessment protocol for both non–engineered organisms. This would help to reduce existing uncertainty for all parties.

Action plan: Set up a scientific review of the current risk assessment, focusing on the appropriate risk questions, levels of certainty and test requirements.

Examine the decisions made under the current approval processes, including:

Were the questions asked appropriate?

Was the amount of information (data) provided to answer the questions appropriate?

Was the correct level of certainty used in reaching the decision?

Was the correct decision reached?

Categorization is very important:

Define a category of uses of microbial control agents that can be adequately regulated under the Plant Pest Act or that can be exempted from FIFRA and Plant Pest Act regulation.

There is no set of universal protocols for assessing the risk of organisms, natural or genetically modified, for use in biological control applications.

FIFRA, EPA and APHIS, USDA have registered or issued permits for the introduction of biological control agents in the environment. The agencies have reached conclusions that the interdiction poses an acceptable level of risk by reviewing data submitted by the applicants. Each organism or agent has been reviewed and the decision announced through registration documentation (FIFRA) or an environmental assessment (APHIS).

Action: Set up an NABC workshop to come up with this categorization

Under current regulations, microbial biocontrol agents, regardless of their intended use, are regulated under FIFRA statutes. Microorganisms, however, are regulated under Plant Pest Act statutes. This distinction is artificial and not related to risk. Requirements for testing and release of microorganisms under FIFRA are not appropriate for many releases or microorganisms since these regulations are intended to certify the efficacy of a product, determine toxicity to non–target organisms as well as assess the ecological risk of releasing the organism. In many cases, such as release of microorganisms that do not produce toxins, or inoculative releases that are not intended for product development, these regulations are too stringent, and create an impediment for testing and release of microbial biocontrol agents, regulation under the Plant Pest Act would be more appropriate for these kinds and uses of microbial biocontrol agents. We recommend that the NABC convene a workshop to develop a proposal for submission to EPA Office of Pesticides and Toxic Substances that describes the uses of microbial biological control agents that: 1 - Can be adequately regulated under the Plant Pest Act, or 2 - are of a class of uses that in a generic sense meet safety requirements and can be exempted from a registration requirements of FIFRA and the Plant Pest Act.

ADOPTION

Leadership and policy are very important

Formulate national and state policy encouraging biocontrol as a preferred means of pest control.

Coordinate activities within federal agencies, between federal agencies and between state and federal levels.

Promote biocontrol as part of a total pest management program.

Identify leadership in each state to facilitate the development and implementation of biocontrol practices. Encourage the identification of a state coordinator for biocontrol.

Biocontrol includes the introduction of exotic agents and the conservation or augmentation of natural enemies.

Regulate biocontrol practices based on an evaluation of the taxonomy of issues relevant to specific biocontrol practices rather than as pesticides.

Aggressively develop biologically based pest control practices

Long–term observations or chemical free cropping systems

Integrate pest control practices in a whole system complex

Use biocontrol as a starting point for whole system pest management

Facilitate large–scale, on–farm trials

Research production, storage, delivery and application methods

Address scientist base, funding adequacy and other incentives supporting specific research objectives

Encourage Extension and farmer participation in appropriate phases of research

Need more basic information on basic biology of pests (life cycle, phenology, environmental controls)

Extension and education is very important

Develop educational materials and make them readily available

Use at–use–site demonstrations

Enhance education for Extension agents and private consultants

Coordinate fact sheets nationally, develop a clearinghouse (biocontrol institute)

Develop Extension programs which facilitate adoption of biocontrols

Need economic assessment

Improve incentives for adoption

Need appropriate economic and environmental assessment of both short– and long– term aspects of all pest control practices

Evaluate biocontrol practices with approaches appropriate to the technology—larger plots, longer term, multiple mortalities and controls

Change cosmetic standards

Communicate value of biocontrol practices

Need public and consumer education effort

Biological control training should be part of pesticides applicator training

BIOLOGICAL CONTROL
MAKING IT WORK

PART 2: TECHNOLOGICAL STATUS

R. JAMES COOK

USDA, ARS Root Disease and Biological Control Research Unit 267 Johnson Hall Washington State University Pullman, WA 99164-6430

ROBERT R. GRANADOS

Plant Protection Program Boyce Thompson Institute Tower Road Ithaca, NY 14853-1801

The use of pesticides for control of insects, weeds, nematodes and plant pathogens includes a worldwide market in excess of $20 billion. However, there are known and/or perceived concerns for such reliance on chemical pesticides for pest control. It is clear that, while chemical pesticide use remains essential to maintain agricultural productivity, new, efficient, and safer products will be necessary to supplement those chemicals that are no longer effective or present a problem to the environment or human health. Moreover, despite this large pesticide usage, it is estimated that nearly 35 percent of the world's food and fiber is still lost as a result of damage by pests and pathogens. Many crop pests and diseases cannot be controlled by pesticides, and many pests once controlled by pesticides have developed resistance to them. Modern technologies rely heavily on the use of chemicals to control these pests.

Biological control strategies offer safe, efficient, and ecologically-based methods for pest and disease control. This report will summarize the current technological status of these agents and identify research needs which will accelerate the use of biologicals as alternatives to chemical control methods, or provide economic controls where none now exist or are acceptable.

CLASSICAL BIOLOGICAL CONTROL

The successful introduction of the vedalia beetle *Rodolia cardinalis* from Australia into California in 1888 to control the cottony-cushion scale

Icerya purchasi, an insect pest of citrus, is considered to mark the beginning of the modern era of biological control of plant pests. The spectacular, low-cost, and permanent success of the vedalia beetle demonstrated the enormous potential of this approach to controlling pests. Indeed, during the last 100 years, there have been over 300 documented examples of complete or significant successes in classical biological insect control. Other outstanding examples of classical biological control in the United States include the control of the olive scale, *Parlatoria oleae*, by the parasites, *Aphytis paramaculicornis* and *Coccophagoides utilis*, during the 1950s, and the control of the walnut aphid, *Chromaphis juglandicola*, by the aphidid parasite, *Trioxys pallidus*, during the 1960s. A recent highly successful biocontrol program in Africa was the use of the encyrtid parasitoid, *Epidinocarsis lopezi*, to control the cassava mealy bug, *Phenacoccus manihot*. Equally spectacular results have been obtained for biological control of weeds as exemplified by the control of Klamath weed by the Klamath beetle, *Chrysolina quadrigemina*, in California. Three other complete successes against weeds since 1944 have resulted in annual benefits of over $150 million per year.

The classical approach of introducing natural enemies to reduce the population of plant pathogens below some economic threshold and in a density-dependent relationship (following the examples from insect and weed control) has met with little or no success for biological control of plant pathogenic fungi, bacteria, viruses or nematodes. The examples of successful classical biological control of insect pests and weeds have nearly always involved the use of an insect and rarely a microbial biocontrol agent and, not surprisingly, classical biological control of one microorganism with another is even more rare. Many studies have been carried out with mycoparasites applied to rust pustules or powdery mildew colonies to limit the amount of innoculum produced at the end of the disease cycle, but there is no commercial application of these kinds of biocontrol agents.

One emerging success story is *Sporidesmium sclerotivorum* applied to soil to keep the inoculum density of *Sclerotina* spp. at or below an economic threshold. This method of biological control keeps the population of *S. minor* in field plots of lettuce in New Jersey in a state of suppression for two or three years following a single initial soil-application of the biocontrol agent. As with so many other promising technologies, the economics of producing a product for control of a single disease or pest tend to discourage companies from investing the capital necessary to produce, formulate, register, and market such a product. However, if used, it might well be the first example of classical biological control using a microorganism applied to soil for control of a soilborne plant pest.

Classical biological control using insects has proven to be an effective and efficient strategy for controlling exotic insect pests and some weeds that have been accidently introduced while escaping their natural insect enemies. It is expected that novel and traditional approaches of ecologically-based biocontrol methods will become even more important in pest management programs, along with genetically improved organisms. However, except for *Drosophila*, there are currently no procedures to engineer the genome of insect species and there is only a limited history of classical genetic modification of natural enemies for biological control of insects with other insects. Determining which characteristics to improve for field efficacy continues to be a major constraint on the genetic improvement of natural enemies. Insecticide resistance, bred into beneficial insects, is a desirable trait since it allows biological control of pests in IPM (Integrated Pest Management) programs that utilize chemical control strategies. Two successful cases of genetic improvement of beneficial insects involve the selection for insecticide resistance in predatory mites in apple (Michigan) and almond orchards (California). As transformation methods are developed in the future, we can expect other genetic alterations of beneficial insects to include host-finding ability, climatic tolerances, changes in host preference, etc.

MICROBIAL BIOCONTROL OF INSECTS

Entomopathogenic microorganisms — including bacteria, viruses, fungi, protozoa, and nematodes — may contribute significantly to regulation of pest insects of importance in agriculture, forestry, and medicine. However, the general lack of success with microorganisms in classical biological control relates in large part to their lack of persistence in the habitat of the target pest agent. This problem has been overcome by applying them repeatedly where needed — in some cases with the same technology used to apply chemical pesticides. It is currently perceived that these entomopathogens have high potential for a greater role in insect pest management by exploitation of naturally occurring entomopathogens as "microbial insecticides." Of the more than 1500 entomopathogenic microbes known, only eleven microbial products are registered by the EPA for use as insect pesticidal agents.

Bacillus thuringiensis (Bt) is the most widely used microbial product and is available worldwide under several brand names. Bt affects more than 150 lepidopterous species, and mixtures of Bt spores and toxin have been used to control caterpillar pests since the early 1960s. Recently, the use has been expanded because of development of improved strains and fermentation techniques that increased production of the bacterial toxin. Bt is currently used extensively for forest and vegetable insect pest control, and its use for stored grain and cotton pests is increasing. *B. thuringiensis* var. *israelensis*, registered in 1981, has been shown to be extremely effective on mosquitoes and black flies. It is used extensively for vector control in North Africa and is gaining greater acceptance for use by mosquito abatement control programs in the U.S. More recently, new Bt strains have been isolated having insecticidal activity against coleopteran (beetle) larvae including the Colorado potato beetle, black vine weevil, boll weevil, and elm leaf beetle. These observations indicate the occurrence in nature of novel Bt strains having activity against a broad array of insect species. This is borne out by the recently reported isolation of nematode-active Bt strains. registration, production, and commercialization of the *Heliothis zea* baculovirus during the 1970s was a major achievement and represented the first viral insecticide to be registered. Three other baculoviruses have since been registered by the EPA and two of these viruses (the gypsy moth and Douglas-fir tussock moth baculoviruses) are commercially available. Four baculoviruses are sold commercially on a small scale in Europe for forestry and agricultural applications. In Brazil, the baculovirus of the velvetbean caterpillar, *Anticarsia gemmatalis,* is used on over 2 million acres of soybean and represents an outstanding example of the use of a baculovirus as a microbial insecticide.

It is estimated that over 750 species of fungi have insect pathogenic effects. To date, only one species, *Hirsutella thompsonii*, is registered in the U.S.; however, this pathogen is no longer commercially available.

Several fungi are registered and have considerable usage overseas. *Beauveria bassiana* is currently in wide use in the Soviet Union against insect pests of orchards and forests. *B. bassiana* is used routinely on several million acres in the People's Republic of China, primarily against European corn borer in corn (maize), Dendrolimus in forest, and plant leafhoppers in rice. In Brazil, *Metarhizium anisopliae* is produced commercially for spittle bug control on more than 400,000 acres of forage and sugarcane. In Europe, *Verticillium lecanii* is commercially available for the control of greenhouse aphids and white flies.

Protozoa use and development has been somewhat more limited than the other microbials. However, one species, *Nosema locustae*, has been developed and registered for use against rangeland grasshoppers. This is the first protozoan receiving U.S. registration.

Entomopathogenic nematodes are being used for agricultural and horticultural insect pest control. Several species in the genera *Steinernema* and *Heterorhabditis* have received the greatest study, and several products are available from commercial sources in North America, Europe, and Australia.

BIOTECHNOLOGICAL ADVANCES WITH INSECT PATHOGENS

In spite of decades of intensive research, relatively few microbial-based products are being used commercially for biological pest management. Clearly, there are major factors constraining the wider use of microbial pesticides and these include lack of environmental persistence, narrow host range, limited virulence, and high production costs. Unlike insect biocontrol agents, for which cloning vectors and other tools for genetic modifications are only now being developed, great advances have been made with the tools of recombinant DNA technology applied to microorganisms, especially bacteria and viruses. Recombinant DNA technology will provide the tools for developing safe, efficient, and cost competitive microbial biocontrol agents. This technology has already made a significant impact on the construction of novel Bt strains which exhibit enhanced insecticidal activity and greater persistence in the environment.

The M-cell delivery system, developed by Mycogen (San Diego, CA), is an example of a novel approach to increase foliar persistence of Bt by a biopackaging process involving microencapsulation of a toxin by the cell wall of a Bt-engineered *Pseudomonas fluorescens* bacterium. Expansion of the host spectrum activity was accomplished by Ecogen (Langhorne, PA) by conjugal transfer of Bt genes and the development of a bifunctional commercial transconjugant product, which has high activity against the Colorado potato beetle *Coleoptera* and the European corn borer *Lepidoptera.*

Both endophytic and epiphytic bacteria, as well as several plant species, have been engineered to express insecticidal Bt proteins. Crop Genetics International (Hanover, MD) is developing endophytic bacteria engineered with Bt toxin genes to protect corn from the European corn borer. Several companies have rapidly developed the technology for introduction and expression of heterologous genes, namely genes from microorganisms including viruses, in plants and have opened new opportunities for the development of insect and disease resistant crops. The utilization of genes encoding lepidopteran- or coleopteran-specific insecticidal proteins from Bt has been used to develop plants (tomato, potato, cotton, etc.) resistant to several insect species under laboratory and field conditions. Recently, Monsanto (St. Louis, MO) developed a novel strategy involving specific modifications of the structural Bt gene sequence that dramatically increased the levels of insecticidal proteins in transgenic cotton plants, and afforded high levels of protection against the cotton bollworm *H. zea* in greenhouse and field tests. In the case of plant diseases, remarkable results are being obtained in virus disease resistance using the coat protein gene of the virus expressed in the plant.

Biotechnological approaches are also being applied to the genetic improvement of baculoviruses. Knowledge of the molecular biology and genetics of baculoviruses has accelerated greatly in recent years, and this database is enabling the genetic manipulation of the viruses. The introduction of foreign genes (Bt toxin, diuretic hormone, and juvenile hormone esterase) into the baculovirus genome which could potentially deliver some deleterious gene product to the insect host has been accomplished, but only marginal insecticidal effects in treated insects have been obtained. It is expected that the engineering of baculovirus genomes with foreign pesticidal genes, possessing enhanced insecticidal activity, will be demonstrated in the very near future.

Studies of the genetics and molecular biology of entomopathogenic fungi, protozoa, and nematodes are still in very early development and it is not likely that recombinant DNA techniques will be utilized to engineer these types of organisms anytime soon.

BIOLOGICAL CONTROL OF WEEDS

Until recently, biological control of weeds was limited to the use mainly of insect herbivores and some foliar pathogens (e.g. rust fungi) released to control specific weeds in perennial ecosystems such as rangeland and pastures. Typical of classical biological control, the success of these agents depends on their ability to multiply and maintain effective populations in response to increases in density of their weed host, and thereby keep the population of the targeted weed in a state of suppression. Such agents tend not to respond fast enough to eliminate a targeted weed within a given window of opportunity typical of most annual cropping systems.

At least four fungal plant pathogens, selected because of their pathogenicity to specific weeds, are now available as "mycoherbicides," and five or six more such products are under development or awaiting registration. The problem of slow response when the epidemic is allowed to develop from entirely natural sources of inoculum is overcome with mycoherbicides by mass production of suitable inoculum in fermentation culture and then applying this inoculum at a time when the weed is particularly vulnerable and temperature and moisture are suitable for infection. The particular pathogens are, in fact, selected because of their limited ability to disseminate naturally, which makes them more ideal as mycoherbicides.

Mycoherbicide products currently available are: DeVine®, a formulation of Phythophthora parasitica for control of milkweed vine in citrus groves in Florida; Collego®, a formulation of *Colletotrichum gloeosporioides* var. *aseschynomene* for control of northern jointvetch in rice and soybean fields in Arkansas, an unnamed product consisting of *Acremonium diospyri* for control of persimmon in pasture land in Oklahoma; and Lubao No. 1, a formulation of *Colletotrichum gloeosporioides* for control of dodder in soybean fields in China. Several others are under development or waiting for registration. Like other microbial biocontrol products, mycoherbicides can be highly efficacious, but the very feature that makes them so attractive ecologically — highly specific for the targeted pest — is the feature that makes them economically unattractive to companies. It is difficult to justify the costs of developing, registering, and marketing a production that controls only one weed, often only one weed in one crop.

BIOLOGICAL CONTROL OF PLANT PARASITIC NEMATODES

Biological control of plant parasitic nematodes in soil by nematode trapping fungi, egg parasites, nematode pathogens, and other natural enemies has been studied for most of this century. Thus far, however, only one biocontrol product has come into the market, a strain of *Paecilomyces* applied to soil for control of nematodes in the Philippines and sold under the trade name Bioact®.

The experience with biological control of nematodes in soil is typical of experiences with biological control of soil-inhabiting insect pests and plant pathogenic fungi in soil: given enough time, e.g., the time provided by a 3-year or longer crop rotation, the naturally occurring biocontrol agents will lower populations of these pest agents below an economic threshold, but elevation or acceleration of these processes above the natural background intensity by deliberate introduction of specific microbial biocontrol agents is generally unsuccessful because of the great difficulty of establishing a microorganism in soil at a population higher than occurs there normally. The successful culture of nematode-susceptible crops without resorting to the traditional long crop rotations, and of replanting some nematode-susceptible orchard crops in the same sites time and again, has been made possible by the use of nematicides, many of which are no longer available or may soon be withdrawn as registration is cancelled or not renewed.

Interestingly, the value or need for crop rotations is now part of a national debate in the United States and elsewhere between advocates of so-called conventional agriculture, and those who practice "alternatives agriculture." Crop rotation, one of the original and still best uses of biological control, has become so underutilized as to be considered an "alternative" practice in agriculture.

At present, the bacterial pathogen *Pasteuria penetrans* is attracting the greatest interest as a candidate microbial biocontrol agent for use against nematodes in soil. There are many strains of this pathogen, some highly specific for their nematode host. Moreover, several cases have now been documented in fields where the natural population of *P. penetrans* accounts, at least in part, for why a certain potentially very important nematode is not important in those fields or region. The limiting factor to use of this agent is the lack of technology for its mass production. When this problem is solved, products of *P. penetrans* can be expected to come onto the market for biological control of nematodes in soil.

BIOLOGICAL CONTROL WITH PLANT-ASSOCIATED MICROORGANISMS

An important breakthrough for biological control of plant diseases is the discovery that on every plant, or within every plant population, there reside microorganisms at a low frequency with the ability to act as antagonists if their numbers were higher to protect that plant or plant population against disease. The evidence suggests further that the frequency of these beneficial microorganisms increases in response to disease outbreaks, but typical of many biological controls, 1– any given population of antagonistic microorganism rarely controls more than one pathogen, and often works only in specific environments, e.g., specific soils, and 2– the antagonistic population of microorganisms tends to build up in response to, rather than in advance of, disease and thus is too late or too slow to control the disease. In spite of these limitations, the prospect exists that the benefits of these microorganisms might be increased significantly by an "inoculative release" with the seed at the time of planting.

The natural occurrence of bacteria represented by *Agrobacterium radiobacter* strain K-84, having the ability to protect plants against crown gall caused by *A. tumefaciens*, and the antibiotic-producing strains of fluorescent *Pseudomonas* species with the ability to protect wheat against take-all caused by *Gaeumannomyces graminis var. tritici* both fit the model of 1– naturally occurring strains of microorganisms on the plant or soil around the plant that 2– increase in numbers in response to the diseases caused by the pathogens they inhibit and 3– provide biological control when introduced in advance of infections by artificial inoculation of the planting material. By inoculating the planting material with these microorganisms, the natural processes involving them are initiated in advance of infection rather than at the end of the disease cycle. Strain K-84 is currently marketed worldwide for biological control of crown gall, and the research on fluorescent *Pseudomonas* strains active against wheat take-all is part of a major research effort worldwide aimed at eventual commercialization of root-associated bacteria for biological control of root pathogens on many crops.

Six microbial biocontrol agents have been or were once registered in the U.S. with the EPA for use against plant diseases. All are plant-associated microorganisms. These are: *Phledia gigantea* (=*Peniophora gigantea*) to control annosus root rot of pines and other conifers; *Trichoderma* applied as BINAB T®to pruning wounds to protect against wood-rotting fungi; *Agrobacterium radiobacter* strain K-84 for use as a bare-root dip to protect fruit tree and rose bush transplants against crown gall; Dagger G®, a peat-based product containing a strain of *Pseudomonas fluorescens* for in-furrow application with cotton seeds to control seedling blights caused by *Rhizoctonia* and *Pythium* spp; *Gliocladium virens* for control of *Pythium* and *Rhizoctonia* on ornamental plants, F-stop®, a *Trichoderma*-based product for use as a biological seed treatment on peas, beans, corn, and other crops to control damping-off. In addition, a strain of *Bacillus subtilus*, sold as Quantum 4000®, is registered for use as a seed treatment of crops, e.g., peanuts in Alabama, because of its plant growth-promoting properties. Plant "growth promotion" is commonly attributable to control of undiagnosed damage caused by subclinical root pathogens.

Plant-associated microorganisms established in the infection court in advance of the pathogen may be used to prevent infection, or as colonists of the infected tissues to arrest disease development. They also may be used to turn on (induce) plant defense genes in the plant. They may even include the pathogen in some avirulent, hypovirulent, or disarmed form (e.g., ice-minus *Pseudomonas syringae*) introduced onto or within the plant to carry out any one of these functions. Plant-associated microorganisms are part of a vast and largely untapped natural biological resource interacting with crop plants. Some, e.g., certain endophytes, have also been shown to control insect pests of plants. Select strains are ready-made biocontrol agents because they are both adapted to the plant or plant part where they must function, and they can be chosen because of their unique ability to protect the plant. Where genetic resistance has not been found, e.g., to the many root- and stem-attacking pathogens, and where soil fumigation is too expensive or not acceptable, epiphytic and endophytic microorganisms are being found with the ability to inhibit these pathogens. Plant-associated microorganisms are another option in plant improvement.

BIOLOGICAL CONTROL OF PLANT DISEASES EFFECTED THROUGH THE PLANT

The greatest progress in biological control of plant diseases during the past 70-80 years has been the approach of improving ability of crop plants to defend themselves, achieved through plant breeding for more resistant cultivars. Some of this effort goes on at a fairly high level of sophistication involving the deployment of specific resistance genes or gene combinations in the plant in response to anticipated or documented (through monitoring) increases in frequency of new virulent biotypes of the pathogen. As expected of biological control, the genetically controlled mechanisms of self-defense in the plant are regulated to turn on when needed, i.e., in response to the pathogen, and where needed, i.e., in the specific cells, tissues, or organs challenged by the pathogen.

The tools to move genes from microorganisms into and express them in plants raises the prospect that many biocontrol mechanisms might be moved into plants, thereby precluding the need to introduce the agent(s). The example of fluorescent pseudomonads active against wheat take-all illustrates both the opportunity and the problems. The genes for phenazine production in the fluorescent *Pseudomonas* species active against take-all have now been cloned and might, therefore, be candidates for transfer to and expression in plants. However, this would require that a root-specific and, preferably, a wound-induced promoter could be used to help insure that the genes would be expressed where the antibiotics are needed, not throughout the plant, and only in response to infection or other root injury, not constitutively. It is also unlikely that one or a few genes from a single strain of acteria could duplicate the multiple mechanisms of disease suppression usually operative with that strain and operative with mixtures of naturally occurring strains responsible for suppression of take-all. Moreover, the use of microorganisms, rather than the plant, to deliver the products of biocontrol genes or biosynthetic pathways can overcome the problem of needing the genes in so many different varieties. There are possibly 30 wheat varieties grown each year across the U.S. in the many areas where take-all and other root diseases are important, and many, if not most, of these are replaced with new varieties every five to ten years. Not all genes need to be in the seed. Some genes may be best deployed in microorganisms delivered with the seed.

BARRIERS TO THE DEVELOPMENT AND USE OF MICROBIAL BIOCONTROL OF INSECTS AND PLANT DISEASES

The success of biological control, including biotechnology-based pest control during the 1990s and beyond will depend on several issues that must be resolved. We must improve the long term, interdisciplinary research on both basic and applied biocontrol problems. Research is needed in areas such as ecology, physiology, and molecular biology to provide the basis of knowledge needed to effectively use biocontrol agents or to genetically improve their pesticidal properties. The transfer of biological control technology from the laboratory to the field is clearly inadequate and must involve a greater enhanced educational program and stronger support of the university extension system. It is imperative that federal and state agencies increase the funds necessary to carry out the research and transfer of technology. In addition, industrial support for some of these programs should be encouraged.

The use of transgenic plants that have been engineered to contain genes which confer pest resistance will have a major commercial impact. However, concerns about the potential development of pest resistance must be addressed by the elaboration and implementation of resistance management strategies. In contrast to this problem, which is real, reference to transgenic plants with ability to express the Bt protein or virus coat protein as "pesticidal plants," or plants with the ability to manufacture their own "chemical pesticides" is not scientifically based and is counter-productive to the advance of this very exciting new approach to biological control of plant pests and diseases.

Repeated examples of natural biological control microorganisms, have revealed a wealth of genetically and taxonomically diverse candidate biocontrol strains. However, the tendency among biological control researchers has been to evaluate the first 10, 25, or possible 50 strains, pick one or two, and then use these as models for studies of mechanisms. We are witnessing an information explosion, mainly because of research on mechanism. It is important to expand the scientific base for biological control, but imagine by analogy a plant breeding program that evaluated only 10, 25 or 50 genetic lines and then ceased to screen lines and concentrated only on mechanisms of disease resistance.

Research programs must be able to both expand our understanding of biological control and continue to examine the virtually unlimited microbial germplasm for new strains or new genes for strain improvement. The ultimate application may take place in the form of microbial biocontrol products, as a new cultural practice favorable to the activity of resident microbial biocontrol agent(s), or both. Such advances, from basic discovery to practical implementation, require long-term commitment by both the researchers and the funding agencies, and they require teamwork and cooperation, including across national boundaries.

Protocols must, and probably can be, developed to both overcome and take advantage of the high degree of specificity of microbial biocontrol agents. We may have no choice but to use different strains or mixtures of strains customized for the disease complexes in specific environments. It should be possible to mass-produce and formulate the individual strains of any given group of microorganisms using a common or standardized technology developed for that taxonomic group, thereby providing some simplification in the use of strain mixtures. Desired mixtures could be produced for the different environments or mixtures of diseases at the local level, possibly even by the user. U.S. farmers already concoct their own mixtures of crop varieties (e.g., wheat) and chemicals according to test results from the experiment stations and advice from the Cooperative Extension Service or local agribusinesses. In a similar but more sophisticated approach to biological control of plant diseases, mixtures of genes are deployed in the same variety or multiline variety and changed periodically according to the ever-changing disease complex and environment.

It is now possible to combine the best traits of several strains into a single strain and possibly preclude the need for complex mixtures of microorganisms. However, scientists taking this approach must be allowed to release their new strains genetically improved by the use of the best tools available. And the system must be in place for the free exchange of microbial germplasm, including across international boundaries, for use in improvement of local strains of microbial biocontrol agents. Imagine where crop production would be if farmers could only grow naturally occurring land races of crops. The current "climate" of unfounded public fears of so-called "genetically engineered microorganisms" and the tendency to over-regulate this technology has discouraged commercial development and use of these engineered products.

Somehow our concept of microbial biocontrol as a component of plant health management must shift away from the chemical paradigm and back to a biological paradigm. Microbial biocontrol agents should be reviewed by the plant quarantine authorities for status as a pest or potential pest, and by the appropriate authorities for safety to people in the workplace or to people or livestock if eaten as a seed treatment, but otherwise every effort should be made to avoid policies based solely on the fact that the biocontrol agent is a microorganism. Microbial biocontrol agents such as mycoparasites, plant-associated microorganisms, and many, if not most, categories of antagonists of plant pathogens or nematodes that have no ability as plant or animal pathogens should be given broad approvals as biological control agents and not be regulated under statutes developed for chemicals. The EPA guidelines for the testing and registration of naturally occurring microbial agents are designed to expedite and facilitate the review of applications. An approach that uses different strains or strain mixtures in different geographic areas, zones, or soils cannot succeed in the context of the existing paradigm, where the regulatory agencies require a separate registration package for each strain or strain combination.

Strain improvement and release programs could be modeled after cultivar improvement and release programs. It would seem likely that companies would eventually take over the work, just as seed companies have eventually taken over the breeding and distribution of some crop cultivars when the economics have been attractive. This model describes almost exactly the sequence of events in the highly successful introduction, and now worldwide use, of strain K-84 of *Agrobacterium radiobacter* for biological control of crown gall.

SELECTED REFERENCES

Adams, P.B. 1990. The potential of mycoparasites for biological control of plant diseases. *Annu. Rev. Phytopathol.* 28: 59-72.

Biotechnology, Biological Pesticides and Novel Plant-Pest Resistance for Insect Pest Management. Proceedings of Conference, Ithaca, NY, July 18-20, 1988. D.W. Roberts & R.R. Granados, eds. 175 pp.

Cook, R.J., and K.F. Baker. 1983. *The Nature and Practice of Biological Control of Plant Pathogens.* APS Press. St. Paul, MN. 539 pp.

Grossman, J. 1989. Update: Biological control of weeds: What's happening, what's needed. *IPM Practitioner* II: 1-11.

National Academy of Sciences. 1987. *Report of the Research Briefing Panel on Biological Control in Managed Ecosystems.* National Academy Press. Washington, D.C. 12 pp.

National Academy of Sciences. 1989. The Ecology of Plant-Associated Microorganisms. National Academy Press. Washington, D.C. 34 pp.

New Directions in Biological Control. R. Baker & P. Dunn, eds. UCLA Symposium Molecular Cellular Biology 1990. New Ser. Vol. 112. Alan Liss, NY.

BIOLOGICAL CONTROL
MAKING IT WORK

PART 3: COMMERCIALIZATION OF BIOLOGICAL CONTROL AGENTS

PAMELA G. MARRONE
President, Entotech
1497 Drew Avenue
Davis, CA 95616

RUEDI SANDMEIER
Vice President
Crop Research
Sandoz Crop Protection
975 California Avenue
Palo Alto, CA 94304-1104

Biological control of unwanted insects, mites, weeds and plant pathogens can be dealt with in two ways. The first is release of a pest's natural enemy into the environment in order to establish itself permanently in the ecosystem and suppress the pest below the economic threshold level. This strategy has been successfully employed in a number of cases and is used mostly for suppression and control of pests recently imported into a new ecosystem. In most cases, this work is being done either by a government agency or contracted by a government agency to private laboratories. Because biological control using natural enemies is usually a one–time event, it will not be further dealt with in this paper.

The second situation is the control of pests with repeated applications of biological agents. This case is mostly used in agriculture, especially annual crops. The annual agricultural system does not easily lend itself to the establishment of permanent population control because the natural fluctuation of the pest population is too high above the economic threshold. The following comments on commercialization refer only to this type of biological control.

WHERE ARE WE?

Although biological pesticides still account for less than one percent of the total pesticide market, sales of biological pesticides are increasing at a rapid pace, estimated to be between 10 to 25 percent per year. Food safety and other environmental concerns, insecticide resistance, and lack of new chemistries have fueled the growth in biological pesticides. Biologists are being integrated into integrated pest management (IPM) systems which are being used in conjunction or in rotation with chemical pesticides.

Microbial pest control agents including plants genetically engineered to produce pesticidal chemicals represent the largest number of recent registrations obtained by the industry. Presently there are 19 microbial pesticides registered with the Environmental Protection Agency (EPA) (Table 1). Since 1984, EPA has reviewed 44 applications for small-scale field testing of genetically modified microbial pesticides. Since 1986, EPA has reviewed 40 applications for small-scale field testing of transgenic pesticidal plants.

Table 1 EPA REGISTERED MICROBIAL PESTICIDES

Microorganism	Year Registered	Target Pest
Bacteria		
Bacillus popilliae/B. lentimorbus	1948	Japanese beetle larvae
B. thuringiensis Berliner	1961	Lepidoptera larvae
Agrobacterium radiobacter	1979	Crown gall disease
B. thuringiensis israelensis	1981	Mosquito/blackfly larvae
B. thuringiensis aizawal	1981	Wax moth larvae
Pseudomonas fluorescens	1988	Pythium/Rhizoctonia
B. thuringiensis tenebrionis	1988	Certain beetle larvae
B. thuringiensis san diego	1988	Certain beetle larvae
B. thuringiensis EG 2348	1989	Gypsy moth larvae
B. thuringiensis EG 2371	1989	Lepidoptera larvae
B. thuringiensis EG 2424	1990	Lep/Coleop larvae
Viruses		
Heliothis nuclear Polyhedrosis Virus (NPV)	1975	Heliothis complex
Tussock moth NPV	1976	Douglas fir tussock moth larvae
Gypsy moth NPV	1978	Gypsy moth larvae
Pine sawfly larvae	1983	Pine sawfly larvae
Fungi		
Phytophthora palmivora	1981	Citrus strangler vine
Colletotrichum gloeosporoides	1982	Northern joint vetch
Trichoderma harzianum/ Trichoderma polysporum	1989	Wood rot
Protozoa		
Nosema locustae	1980	Grasshoppers

Bacteria—Although more than 100 bacterial species have been identified as insect pathogens, only certain Bacillus species have enjoyed commercial success.

Of these, *B. thuringiensis* has been most widely exploited. In addition, *B. popilliae* has been sold for many years to home gardeners to control Japanese beetles. *B. sphaericus* has been developed into a commercial product for mosquito control by Abbott Laboratories.

Table 2 summarizes commercial Bt based products. Never before has there been such interest by industry in Bt as evidenced by the proliferation of start–up companies (Ecogen, Mycogen, AGC, PGS, etc.) and large agro-chemical companies developing Bt based products. However, Bt based products still have a number of constraints, which the various companies are addressing through a number of strategies.

Table 2 **COMMERCIAL *B. THURINGIENSIS* BASED PEST CONTROL AGENTS**

Organism	Target Pests	Commercial Products	Company
B. thuringiensis			
var. *kurstaki*	caterpillars	Dipel	Abbott
		Thuricide/Javelin	Sandoz
		Biobit/Foray	Novo Nordisk
		Bactospeine/ Futura	Duphar
		Cutlass/Condor	Ecogen
		MVP	Mycogen
		Agree	Ciba–Geigy
		Wormbuster	Bactec
var. *israelensis*	Mosquitoes/ Blackflies	Vectobac	Abbott
		Teknar	Sandoz
		Skeetal	Novo Nordisk
		Bactimos	Duphar
	Fungus Gnats	Gnatrol	Abbott
var *tenebrionis*	Colorado Potato Beetle	Novodor	Novo Nordisk
		Trident	Sandoz
		Foil	Ecogen
var. *san diego*		M–One/ M–One Plus	Mycogen
var. *aizawai*	waxmoth	Certan	Sandoz

Narrow host range—Although the narrow host range is seen as an advantage economically, it often limits the market potential for Bt products. To address this issue, Ecogen developed Foil® a product for controlling European corn borer, a Lepidopterous pest and Colorado potato beetle, a Cleopteran pest on potatoes. Through plasmid curing and transfer, a 150 Md plasmid from a *kurstaki* strain and an 88Md plasmid from a *tenebrionis* strain were combined into one organism.

Ecogen, Ciba–Geigy and Sandoz–Repligen scientists have used various molecular techniques (electroporation, transconjugation, etc.) to develop products combining genes to increase the activity on key Lepidopteran pests such as armyworm *Spodoptera* and cotton bollworm *Heliothis spp.* In addition, fusions of baculovirus with Bt *tenebrionis* genes and Bt *israelensis* with Bt *kurstaki* genes are being used to expand the host range. These products are in various stages of development and commercialization.

Several companies have amassed large collections of Bt isolates from soil, plants, insects, grain dust and other environmental niches, with the objective of finding strains with enhanced activity on key pests and new activity on pests such as corn rootworm. After bioassays of 7000 isolates against *Diabrotica undecimpunctata*, Ecogen reported to have discovered a weakly active strain. ICI (Imperial Chemical Industries) also reported a similar discovery.

Poor residual activity—It is well known that Bt lasts only a few days on plant foliage under typical field conditions due to UV degradation, rainfall, etc. Mycogen uses *Pseudomonas* as a delivery system for Bt genes (*kurstaki* for MVP®and *san diego* for M–One Plus®). The Bt bearing *Pseudomonas* is killed (to avoid regulatory hurdles for registering recombinant microorganisms) and sprayed on the crop like other Bt products. The Pseudomonad wall is reported to protect the Bt protein from environmental degradation, thus providing longer residual activity.

The USDA research laboratory in Peoria, Illinois has developed a starch encapsulation method for protecting Bt and viruses in the field. Field trials of experimental products are being conducted.

Novo Nordisk's Foray®for gypsy moth control has a unique formulation designed specifically for aerial applications to forest canopies. The formulation has unique rainfast, palatibility, and settling properties.

Inconsistent control on cryptic insects—Because Bt must be ingested to act on the insect, there are a number of fruit feeding insects that are difficult to kill under field situations. For example, the cotton bollworm takes only a few bites of cotton tissues before entering the square. A number of companies have chosen to engineer Bt gene(s) directly into the plant or into plant colonizing microorganisms to address this problem. Monsanto, PGS, Agracetus, ICI, Sandoz, Calgene and Ciba–Geigy all have plant genetic engineering programs for insect control.

Crop Genetics International has transferred a Bt gene into an endophyte, *Clavlbacter xyli*, for control of European corn borer, a stem and ear feeding pest. Because of the difficulty in controlling this insect and low cost of corn, insecticides are not widely used on this insect. CGI aims to tap into this market potential with InCide®, one of the few recombinant microorganisms approved for outdoor field tests.

Although there could be a number of disadvantages to plant genetic engineering Bt genes, such as resistance development and public acceptance of engineered foods, companies see several advantages in addition to increased efficacy. These are increased potential for a proprietary position, improved residual activity, and reduction in chemical pesticides. In addition, field testing transgenic plants, which is regulated through APHIS, has been considerably easier than field testing transgenic microorganisms through the EPA.

Weak intellectual property protection—An additional constraint to the commercialization of Bt and all other biological agents is the state of our intellectual property protection. Biological control in the future, if it is to go beyond the small niche it occupies today, will need investment by private industry and government agencies. Private industry is only willing to make those investments if it can enjoy protection from competitors granted through the patent process. However, while the patent process is very clear and very well understood for chemical pest control agents, it is not clear for biological agents, and a large backlog in processing biotechnology patent applications has not helped industry gain the confidence that it will have a guaranteed return on investment.

VIRUSES

There are currently five insect viruses registered in the United States (Table 1). Commercialization of viruses has lagged behind Bt primarily because of the lack of an economical *in vitro* production system. Large agrochemical companies (with the exception of Sandoz's former entry into Elcar®production) have avoided developing *in vivo* rearing production systems. Progress being made in *in vitro* production system is largely being driven forward by pharmaceutical applications for baculovirus expression of proteins. One company, American Cyanamid, is developing *in vitro* production for an insect (gypsy moth) virus.

Espro, a small company based in Maryland, is producing GypChek®for the U.S. Forest Service. Espro has developed an economical *in vivo* production system for this virus, and hopes to expand into other viruses fro control of codling moth, *Heliothis*, and *Spodoptera*.

Lou Falcon, from the University of California at Berkeley, is producing codling moth virus for sale under the Experimental Use Permit (EUP) to fruit growers in California and Washington. Registration is expected in 1991, and the virus will then be available for licensing to a company for large scale commercialization. This virus is already registered and produced in a number of European countries.

In addition to a lack of large–scale economical *in vitro* production methods, insect viruses have many of the commercial weaknesses as Bt. Baculoviruses have an even narrower host spectrum and take longer to kill than Bt. Like Bt, they must be ingested and also rapidly degrade on plant foliage. A number of approaches are being developed to address these weaknesses, with much of this research being conducted at universities. Several groups are exploring insertion of insecticidal protein genes (Bt, juvenile hormone, insect specific neuropeptides) into the viral genome in order to reduce the time to kill the insect. In addition, gene deletions (Louis Miller, University of Georgia) and viral proteins (Robert Granados, Boyce Thompson Institute) have been shown to increase the unit activity of certain viruses. Researchers at the Boyce Thompson Institute performed the first outdoor field test of a genetically altered virus.

FUNGI

Fungi have been developed into very few commercial products. Although many fungi are known to be very effective against insects in laboratory tests, efficacy in field situations has not met expectations. To address the problem, EcoScience, a Massachusetts based company, has developed bait and trapping techniques for delivering well-known, but effective insect pathogens such as *Metarhizium* and *Beauveria* against flies and cockroaches. Products based on these bait traps should be available in the early 1900s.

Fungi for agriculture have not been exploited due to lack of formulation and delivery technology that provides protection against desiccation, wind, rain, UV light, etc.

PROTOZOA

The only commercial product based on protozoans is *Nosema locustae* for grasshopper control. Protozoa are not viewed by the industry as having commercial potential because of the chronic nature of the activity and requirement for *in vivo* production.

ENTOMOGENOUS NEMATODES

Two companies have invested considerable resources into development of nematode based products. These are Biosys (Palo Alto, CA) and BioEnterprises (Hobart, Tasmania). Biosys sells BioSafe®, a home and garden product for catalog sales and for cranberry weevil control and BioVector®, for citrus weevil control. A grub product is in development with Ciba–Geigy. Biosys markets *Steinernema* spp. produced in liquid fermentation at a fermentation plant in Alberta Canada.

BioEnterprises, owned by Biotechnology Australia and Hoechst sells Heterohabdltid nematodes produced in solid culture for home and garden and nursery applications in Australia and Europe. Products for U.S. turf markets are being developed in conjunction with ChemLawn.

Development of nematode products has been slow, although considerable progress in commercialization has occurred in the last five years. Limitations continue to be economic production, shelf life and quality control.

FUNGICIDES

A number of companies were developing biofungicides in the early to mid-1980s. Much of the development centered around plant colonizing bacteria, such as *Pseudomonas*, which could be engineered to improve fungicidal activity, contain new pesticidal genes or improve plant colonizing characteristics.

Very few companies are still involved in this type of biofungicide research and development. Expectations for the technology were higher than could be delivered in the short time frame planned for commercial development of the products. In addition, recombinant microbial pesticides encountered considerable regulatory and public acceptance hurdles (ice–minus, for example) making continued development too costly for many companies.

Despite the setbacks, there is a core of university and USDA researchers that are pushing forward the biofungicide area. In addition, W. R. Grace, Kodak and Gustafson have commercialized biofungicides for greenhouse use and as seed treatments.

HERBICIDES

Sandoz, Mycogen and Crop Genetics International have active programs in bioherbicide development. Dow–Elanco is commercializing bioherbicide products in Canada. Bioherbicides have not had as much attention as bioinsecticides because of the availability of inexpensive, relatively safe and very effective chemical herbicides. This will continue to be the situation until the market situation changes.

WHERE ARE WE GOING?

Biological pesticides will continue to increase in importance relative to chemical pesticides, but will not replace the need for chemical pesticides. Biological insecticides will remain the most important segment, followed by biofungicides and then bioherbicides.

Biological pesticides have been increasing in efficacy while becoming competitive in price with chemical pesticides. This trend will continue, increasing the grower acceptance of biological products. In the next decade, we will see a mix of naturally occurring, genetically modified (developed through conjugation, plasmid curing, classical mutation, etc.), genetically biologicals, transgenic pest resistant plants and chemicals integrated in pest management systems.

HOW CAN WE GET THERE?

For biologicals, whether naturally occurring or engineered, to reach their full potential, there are some key areas for continued focus:

1—Clear, consistent federal regulation of microbial pesticides in the existing federal regulatory framework.

2—Regulatory focus on the pesticide product, not the process by which it was produced.

3—State legislation consistent with federal regulation.

4—Public understanding and acceptance of products developed through non–traditional methods (recombinant DNA techniques).

5—Adequate state and federal funding for basic university research (mechanism of action of Bt and viruses, physiology of entomogenous nematodes, Bt resistance monitoring and management strategies).

6—Economic incentives for development of scale up technologies for insect viruses, predators, parasites, fungi and nematodes in order to make the products more attractive to industry.

7—Increased research and development of formulation and application technology of microbial products.

8—Increased grower education of the differences between chemicals and biological products (necessity for increased scouting, timing, attention to application, etc.).

9—Clear intellectual property protection policy by the Patent Office, expedited treatment of patent applications, and a guaranteed lifetime of those patents in a way that industry can recoup its money.

Note: Points 1–3 and 6 are addressed in the President's Council on Competitiveness: Report on National Biotechnology Policy (February, 1991). This is a very positive sign for the future commercialization of microbial and recombinant plant products.

BIOLOGICAL CONTROL
MAKING IT WORK

PART 4: REGULATION OF BIOLOGICAL CONTROL AGENTS

The United States government essentially regulates biological control agents under statutes enforced by the Environmental Protection Agency (EPA) and the Animal and Plant Health Inspection Service (APHIS) of the U. S. Department of Agriculture (USDA). The intent of this paper is to give a broad overview of these regulations only. Details should be obtained directly from the proper agency. Emphasis is placed on discussion of the current practice of viewing genetically modified organisms differently, although the policy of the Federal Government is that regulation should be by the product, not the process by which it was developed.

SUE ANN TOLIN

Department of Plant Pathology, Physiology and Weed Science
Virginia Polytechnic Inst.
Blacksburg, VA 24061

CURRENT REGULATION BY EPA AND USDA

Microbial pest control agents are currently regulated by EPA, since provisions of the Federal Insecticide, Fungicide, and Rodenticide Act (FIFRA) give EPA authority over the distribution, sale and use of pesticide products. The definition of pesticide used in FIFRA is: 1– any substance or mixture of substances intended for preventing, destroying, repressing, or mitigating any pest, and 2– any substance or mixture of substances intended for use as a plant regulator, defoliant, or desiccant, except for animal drugs and animal feed containing an animal drug. The latter items come under the USDA and the Virus Serum Toxin Act.

Registration under FIFRA is required for all pesticides, chemical or biological. The registration is qualified by specifying conditions for approved use and distribution. Registration approval is based on data that allow EPA to conclude that the product, when it is used in accordance with the specified conditions, will not cause (or significantly increase the risk of) unreasonable adverse effects to humans or the environment. Investigators must obtain the data to support the conclusion, including data in the areas of 1–product analysis, 2–residue, 3–toxicology, and 4–non-target organism and environmental expression, according to end-use pattern and the test substance.

Initial testing of a pesticide for efficacy, and tests to obtain required data for registration may be done under an experimental use permit (EUP) under FIFRA. Traditionally, EPA has maintained that small-scale tests are generally tightly controlled, involve small quantities of pesticides, and are conducted by highly-trained personnel, and thus pose minimal hazards. Thus, EUPs are not generally needed for tests such as those conducted on under ten acres of land, if the food or feed crops involved are destroyed or fall within an approved tolerance level. Microbial pesticides were included in such determinations, but were not automatically exempt. The Agency retained the authority, however, to require an EUP if it determines that EPA oversight is warranted. In addition, APHIS review might be needed if the organisms may be a plant pest.

Consideration of the issues surrounding small-scale tests of genetically modified or nonindigenous microbial pesticides led EPA to conclude in 1986 (51 *Federal Register*, 23302) that its new policy would be to require an EUP because it was judged that oversight was warranted. This was based on the presumption that even small-scale tests with genetically modified organisms may raise many of the same concerns as more extensive uses of conventional pesticides. In its policy statement, EPA requested that investigators notify EPA before starting any field testing with such pesticides so any small-scale test could be evaluated for possible risk to human health or the environment.

The notification allows EPA to make a determination as to whether an EUP is required. If it is, EPA issues the EUP; if not, the investigator has the EPA blessing to proceed. The purpose of the EUP, as originally stated by EPA in 1974 (39 *Federal Register*, 11306), is to strike a balance between facilitating—or, as a minimum, not unduly impeding—pesticide research and development and protecting against human and environmental injury. The statement recognized that hazards may be posed, yet acknowledged that experimental use and testing are essential to the development of new, less hazardous, more effective pesticides, including both chemical and biological agents.

EPA is currently drafting regulations for implementation of this policy. Comments were requested in February 1989 (Federal Register 54:7027) on the regulatory approach to biotechnology under FIFRA. It is expected that the new regulations will spell out details of the data requirements, and clarify issues relating to scope of coverage.

USDA-APHIS has traditionally provided oversight of introduction of organisms, such as insects, used in biological control under the Plant Quarantine Act and Federal Plant Pest Act. These allow APHIS to regulate the movement into and through the United States of plants, plant products, plant pests, and any product or article which may be or contain a plant pest at the time of movement. A plant pest means "any living stage of any insects, mites, nematodes, slugs, snails, protozoa, or other invertebrate animals, bacteria, fungi, other parasitic plants or reproductive parts thereof; viruses, or any organisms similar to or allied with any of the foregoing; or any infectious substances which can directly or indirectly injure or cause disease or damage in any plants or parts thereof; or any processed, manufactured, or other products of plants."

Additional rules were promulgated by USDA-APHIS in 1987 (7 CFR 340) to cover the introduction of organisms and products produced through genetic engineering (i. e., organisms modified by the use of recombinant DNA techniques) and which contain or use plant pests in their construction. Permits are currently issued by APHIS for the "introduction" (to move into or through the United States, to release into the environment, or to move interstate) of "regulated articles" (any organism altered or produced through recombinant DNA techniques if the donor organism, recipient organism, or vector or vector agent belongs to a genus or taxon that meets the definition of a plant pest, or is an organism of unknown classification).

Field tests with genetically modified plants have been approved by USDA-APHIS under this mechanism. APHIS permits are not required, however, for comparable field tests with organisms, i. e., plants, produced by methods which do not involve recombinant DNA techniques.

PROBLEMATIC ISSUES IN THE REGULATION OF BIOLOGICAL CONTROL AGENTS

Several issues relate to the overlapping authorities of Federal Agencies in the regulation of biological control agents, particularly if they are genetically modified. Although Agencies have been interacting quite nicely in the initial phases of research, several issues will need to be resolved before applications become widely used.

Some biological control agents, because of their end use characteristics or components, are subject to both EPA and USDA regulations. For example, *Clavibacter xyli* subsp. *cyanodontis* tested for control of the corn ear worm was regulated by APHIS because it has plant pest properties. Additionally, EPA is considering regulating plants modified to have resistance to plant pests as pesticides, and thus such plants could be called biological control agents. This concept markedly differs from traditional practice. If the end use of plants containing either of the above examples is used eventually as human food, the Food and Drug Administration (FDA) will become involved in the registration of biological control agents.

There is currently no mechanism in Federal regulations to obtain permits for research trials that do not envision a pesticidal product in the near future. The EPA regulations apply primarily to the development of a pesticide for registration and eventual commercial application. This poses a potential problem in the "passing of the baton" between the Agencies for certain tests.

Envision a scenario in which a permit was obtained from EPA to conduct small-scale field tests with genetically modified bacteria, involving collaborative arrangements between a company and an academic researcher. After the first year of the test, the company deems the efficacy too low to proceed and terminates the test and the relationship. Meanwhile, the researcher would like to continue observation of the site and/or repeat the test with different parameters to ask scientific questions relative to why the test did not work. However, since the test is not being conducted to obtain pesticide data, EPA is reluctant to provide an EUP for the test. The researcher is reluctant to proceed without any type of review.

WHERE DO WE GO FROM HERE?

The United States has been widely complemented for progressing as rapidly as it has in the regulation of genetically modified organisms. Continued efforts need to be made, particularly with bacteria and other microorganisms, where the greatest promise as biological control agents exists. A sound, but scientifically sensible, regulatory system is essential for making biocontrol work.

Scientists need to be encouraged to apply the most powerful techniques to improve microorganisms, e. g. molecular methods, to achieve better activity as biological control agents. The current regulatory climate, however, has not provided that encouragement. On the contrary, the fact that investments by the private sector have been decreasing has been attributed in part to the regulatory climate. The situation must improve before we can go forward.

BIOLOGICAL CONTROL
MAKING IT WORK

PART 5: SUPPORTING ADOPTION OF BIOLOGICAL CONTROL

CAROL S. GLENISTER
President
IPM Laboratories
Main Street
Locke, NY 13092

If we divide the practice of biological control into two parts, microbes and macrobes, we find two very different states of development. The microbes, such as Bt, are well-known as commercial entities, although relatively few in number when compared to the number of chemical pesticides. Whereas most growers know what *Bacillus thuringiensis* is, growers are just becoming aware of the existence and use of macrobes that we call natural enemies such as crytolaemus, Encarsia formosa, lacewings, trichogramma, and beneficial insect–parasitic nematodes.
This paper is based on the experience of members of the Association of Natural Biological Control Producers (ANBP). At this point, members of the ANBP are primarily producers of natural enemies. However, our challenges and needs are paralleled closely in the microbial industry. Since microbials are well-known in comparison to natural enemies, I will concentrate on examples from the lesser-known.

CURRENT STATUS

Commercial use of beneficial insects and mites almost disappeared after World War II, when biological control retrenched to classical exploration and importation of natural enemies of exotic pests. Production and releases of native natural enemies in the United States has made a gradual reappearance in mainstream agriculture over the last three decades. Today we find natural enemy production and demand accelerating out of the lag phase toward the log phase of the sigmoid curve that describes population growth. Major users include greenhouses, farms, and grower cooperatives. Public and research growing facilities nationwide are rapidly adopting parasites and predators as well.

In the United States today, planned releases of spider mite predators are made annually on approximately 75 percent of California's 20,000 acres of strawberries. Releases of trichogramma egg parasites for worm control on cotton cover a conservatively estimated minimum of 50,000 acres. Release of predators and parasites against mealybug and scale are made on over 28,000 acres of California citrus. Lacewings are applied on over 10,000 acres both from the air and ground for control of aphids and other pests on many crops. Beneficial insects and beneficial mites are used in over 90 percent of British Columbia's vegetable greenhouses.

THE FUTURE FOR NATURAL ENEMIES

ANBP's goal (see p. 249), is to promote the production and supply of effective natural enemies of high quality for the management of pests. As regular witnesses of effective pest management through use of natural enemies, we envision the day when such biological controls are used regularly throughout all of agriculture. Just as pesticide users had to learn what type of nozzles to use, the most effective spray pressure, and the minimum effective spray volumes to get good chemical control, quite a lot of learning is ahead for the new biological control users. But it will be done. Pesticide use will also continue, but pesticides will be managed so that they have minimal impact on natural enemies. More careful pesticide management will grow together with pest managers growing awareness of natural enemies.

MAJOR HURDLES

Our major hurdles include general pest control habits, pesticide research techniques, regulatory habits, large information gaps, and the development of systems of massive quality production.

General Pest Control Habits—Essential for successful biological control are:

- –pest monitoring
- –creation of an environment friendly to beneficial pesticide management
- –releases of most natural enemies when the pest levels are still extremely low

These needs all come under a strategy familiar to us: Integrated Pest Management (IPM). Continued development of IPM and IPM professionals throughout agriculture will make it possible for biological controls to be effectively used.

The single greatest threat to continuous dedicated use of IPM is the silver bullet: the fantastic new pesticides or engineered plant that annihilates multiple pests with a single application for the entire season. Silver bullets encourage pest managers to stop monitoring because there is nothing to look for. Natural enemies are either killed or have nothing at all to eat (eventually resulting in secondary pest outbreaks). And massive pest resurgence resulting from pest resistance and absence of natural enemies almost always ensues in a few years.

For these reasons, we recommend that public funds support research and development only for pest controls that are friendly to natural enemies. We need to be cautious in setting up barriers to new, broad–spectrum pesticides, however. For example, although the broad spectrum pesticide, Avid, is extremely effective against many types of pests, its permissiveness to natural enemy survival makes it an extremely useful IPM tool for knocking back outbreaks before applying natural enemies.

Pesticide Research Habits—The established routine for testing pesticides is to set out randomized treatments replicated over small plots to test for significant differences among treatments. The experience of ANBP producers is that small plot trials are often inadequate for evaluating natural enemy species. Some of the problems encountered are the natural enemies' movement and cross-contamination of the small plots, including their demise in the pesticide plots. One example is the very carefully done 1960s small plot trial that that demonstrated biological control of spider mites in strawberries (Oatman, *et al.*, 1968). The trial clearly demonstrated that exorbitantly high numbers of predatory mites were needed for acceptable spider mite control. According to Glen Scriven, co–founder of BioTactics, this knowledge put a tremendous damper on this spider mite control technique. One pest control management consultant, Charles Wood from the Oxnard area persisted in predator releases for nearly twenty years until Driscoll's, a California strawberry growing cooperative, began to research predators in the early 1980s. By 1988, they faced a desperate situation: they expected the miticide Plictran®to be withdrawn from the market, and spider mites were exhibiting resistance to the remaining miticides. In 1988 Driscoll's began large scale trials with the predatory mites. Over a period of about three years, backed by Wood's and Driscoll's experience, growers turned en masse to predatory mite releases.

Today, large scale experience indicates that a release rate of 20,000 predatory mites per acres very early in the season is usually adequate for spider mite control.

The biological control experience in strawberries demonstrates the two most basic needs for biological control adoption.

1–a belief that biological controls can keep pests at acceptable levels.

2–a determination to figure out what is needed to make biological controls work.

The need for large scale trials rules out the possibility of generating rigorous statistical data. Instead, researchers need to rely more on experiential, unreplicated observation that is not generally accepted in the research community as "scientific."

Regulatory Habits—Regulations are impeding expansion of biological control by:

1–Setting up barriers to the movement of natural enemies from state to state. Natural enemy providers must deal with the individual regulations of each state to which they ship arthropods. For example, this month the Florida agriculture department sent us letters requesting that we ship them samples every six months of the natural enemies that we are sending over their borders. Such samples are a good idea, however, it will be expensive for us producers to replicate samples semi–annually for individual states.

2–Setting up barriers to the registration of new pesticides that are more compatible with natural enemies. For example, the plant extract neem has been registered as a pesticide in several states, but the registration process in New York state is so cumbersome that we will be extremely surprised if neem becomes commercially available in New York state in less than two years.

Information Gaps—The greatest initial need of every biological control practitioner is clear, concise fact sheets on individual natural enemy species, their life cycle, environmental tolerances (pH, temperature) and ranges in their developmental rates, feeding rates, and reproduction rates. Readily available facts would greatly reduce the guesswork in deciding when, where, and how to apply. Guidelines on application rates, methods, coverage, timing, and the reasoning behind these would greatly enhance the rate of user success. Updated information on relative compatibility of individual natural enemies with different pesticides is badly needed.

Rather than have each state extension service publish individual fact sheets, it would be more efficient to do the fact sheets on a national level, with regional emphasis placed on regional demonstrations and field observations. Large scale trials are needed to test guidelines, recognize weak areas, and prioritize further study needs. Emphasis should be in the ultimate objective, the large–field laboratory with field observations feeding back to the scientific laboratory for refined testing on hypotheses of why given things are happening.

Massive Quality Production—All of the above depends upon the massive supply of healthy natural enemies at a reasonable cost. Constant improvements in production efficiency, quality controls, storage, and shipping will come both from the private and public sectors.

PARALLEL NEEDS STATED BY RESEARCHERS

Biological control researchers have stated similar and additional needs enumerated below:

In 1979, Ridgway, et al. (1981) identified trichogramma research needs general enough to apply to the entire realm of natural enemies. These are paraphrased below:

1–Identification and selection of most effective species of strains.

2–Efficiency improvements in production and release.

3–Quantitative observations and analysis of the impact of releases on pest populations and crop yields/quality.

4–Improved methods of pest monitoring and pest forecasting.

5–Management of pesticides to minimize impact on natural enemies.

During that same conference, Tauber and Helgesen (1981) identified the leading impediments to the use of biological controls in the United States greenhouses as: 1–lack of supply; 2–lack of experience and basic knowledge, and 3–lack of a sales and support system like that found for chemical or seed sales.

Private and public sectors need to cooperate in addressing these factors.

SELECTED REFERENCES

Oatman, E. R., J. A. McMurtry, and V. Voth. 1988. Suppression of the two–spotted spider mite on strawberry with mass releases of Phytoseiulus persimilis. *J. Econ. Entomology* 61:1517–1521.

Ridgway, R. L., J. R. Ables, C. Goodpasture, and A. W. Hartstack. 1981. Trichogramma and its utilization for crop protection in the U.S.A. J. R. Coulson ed. In proceedings of the joint American–Soviet conference on use of beneficial organisms in the control of crop pests. Washington, D.C., August 13–14, 1979. Entomological Society of America, College Park, MD.

Tauber, M. J. and R. G. Helgesen, 1981. Development of biological control systems for greenhouse crop production in the U.S.A. J. R. Coulson ed. In proceedings of the joint American–Soviet conference on use of beneficial organisms in the control of crop pests. Washington, D.C., August 13–14, 1979. Entomological Society of America, College Park, MD.

ASSOCIATION OF NATURAL BIOCONTROL PRODUCERS (ANBP)

Critical Issues

Regulation of Biological Control Organisms—The ANBP is a non-profit organization representing the producers and distributors of natural enemies, including predators and parasites. Our goals are as follows:

1—Provide a unified approach for the promotion and improvement of the natural enemy industry.

2—Foster programs which enhance the quality and standardization of natural enemy production.

3—Encourage self regulation to enhance professionalism of its membership.

4—Work for the continued improvement in the quality of agriculture production as well as enhancement of the environment.

5—Promote research and education on the use of natural enemies.

Critical issues affecting the natural enemy producers can be divided into two groups, government input and producer input. The goal of both producer and government activities is to provide the public with a quality supply of effective natural enemies for the management of pest species.

Regulation (state and federal)—

1—An extremely important issue for us is the creation of regulations of the natural enemy production industry that are distinct from pesticide regulations. We also consider it important to be considered a separate industry from the production of genetically engineered biotechnology organisms. Due to chemical toxicity and environmental risk both of these groups require EPA regulation. We believe that since we are rearing animals that already exist naturally in the environment and pose no threat to public health, stringent federal registration and regulation are unnecessary.

2—In order to optimize the effectiveness of reared natural enemies in agriculture, licensing of producers and training and licensing of biological control practitioners is desirable. This licensing and training should be distinct from traditional pesticide licensing and training programs.

Research (state and federal)—For many years the focus of entomological research has focused on the efficacy of pesticides. The time has come to direct research funds to optimizing the use of natural enemies. Grants should be awarded to researchers to find more effective natural enemies, develop advanced production systems for commercial insectaries and conduct large scale field trials to determine the practicality of natural enemy use.

Information and Education (state and federal)—The lack of information on the effective use of natural enemies is a major impediment to the expansion of the use of natural enemies. Access to federal databases, the compilation of research literature and the publication of reliable information in a form usable by the public is essential.

Producer Responsibilities—

1—Marketing of natural enemies:

The producer and distributor must make every effort to provide accurate information to the user of the natural enemies. The method of holding and release of the natural enemies is critical to their successful use. Guidelines for the release rates of natural enemies by crop and pest species are needed.

2—Identification of beneficial organisms:

The producer will collect samples of cultures on a regular basis and submit them to a designated professional taxonomist for identification.

3—Quality control:

Develop practical quality control techniques that can be incorporated into existing production systems. Develop standardized sampling techniques to maintain consistent quality and numbers of natural enemies shipped.

4—Code of Ethics:

Members of ANBP subscribe to a code of ethics regarding the use of beneficials.

The use of applied biological control (mass release of natural enemies) can be an important component in the management of many pests. As additional pesticides are removed from the market, alternatives must be found that are safe for the public and the environment. The ANBP and its members are ready to work with government agencies to develop the tactics for mass release of natural enemies in an overall strategy of integrated pest management.

THE DEVELOPMENT OF SHEEP EXPRESSING GROWTH PROMOTING TRANSGENES

JAMES D. MURRAY
Animal Science
University of California,
Davis, CA 95616

CAIRD E. REXROAD JR.
USDA ARS Barc East
Building 200, Room 13
10300 Baltimore Avenue
Beltsville, MD 20705

The first successful attempts to transfer foreign DNA into mice (Gordone, *et al.*, 1980) were rapidly followed by a number of reports demonstrating that foreign DNA, or transgenes, can be permanently incorporated into an animal's genome (Costantini and Lacy, 1981; E. Wagner, *et al.*, 1981; T. Wagner, *et al.*, 1981). Once incorporated, transgenes are stable and can be transmitted to an animal's progeny in the course of normal breeding.

The potential for the application of transgenic technology to alter animals used in agriculture was further indicated by Palmiter and his colleagues (1982, 1983). These workers produced mice transgenic with a rat growth hormone (rGH) gene that was under the control of the mouse metallothionein I (MT) gene promoter, thus directing the production of GH in the liver rather than the anterior pituitary. The resulting transgenic mice that expressed this transgene grew to approximately 1.5 to 2 times as large as non-transgenic littermates. Thus, they established that the alteration of the pattern of expression of a gene, or the expression of a foreign gene, could result in a marked change in a quantitative trait like body growth. The control of growth, feed efficiency, reproduction, fat deposition and disease resistance are all economically important traits in animal agriculture.

At this point in 1982, laboratories at the USDA/ARS facility at Beltsville, Maryland (C.E.R. Jr.) and the CSIRO, Division of Animal Production in Sydney, Australia (J.D.M.) began experiments designed to transfer growth hormone transgenes into sheep. The goals of both groups were to determine 1–if growth hormone transgenic sheep were more feed efficient, i.e., produced more muscle per unit of food consumed, 2–contained less fat and 3–grew more rapidly than non-transgenic controls, as such modifications would be advantageous to producers. In this paper we will discuss the experimental work concerned with producing growth hormone transgenic sheep leading to the present, as well as our thoughts on the ethical and social considerations of this work.

ETHICAL IMPLICATIONS

Should we use genetic engineering? A frequently asked question is "Do we have the right to change the genetics of a species?" and each member of these research teams had to answer that question to his, or her, own satisfaction. As agricultural and basic research scientists, we realized that this question had already been answered thousands of years ago. Soon after mankind first domesticated plants and animals selective breeding began to be practiced, thus leading to man-made changes in the genomes of all domesticated species.

Wild cattle were bred to become Herefords, Angus and Holsteins. The original domesticated wild dogs have been selected for different purposes and bred until today there are a multitude of breeds. All of our domesticated animals, both those used in agriculture and the companion species, are now represented by a wide variety of breeds or types. The same is true for plants, although for plants mankind has gone much further. Using currently acceptable breeding techniques plant breeders have already produced entirely new species, such as the grain triticale, for use in agriculture.

Traditional breeding techniques rely on hybridization between breeds or species to yield genetic variation upon which we can impose selective breeding to fix desirable gene combinations. Essentially this involves manipulating and introducing changes into an animal's entire genome, containing perhaps as many as a hundred thousand genes. By comparison, genetic engineering involves introducing one or two characterized genes into one animal and assessing the effect of the expression of that gene on that animal and its immediate progeny. Thus, to us, transgenic technology is a more precise extension of the genetic manipulations characterized by selective breeding which farmers and agricultural scientists commonly practice.

TECHNICAL DEVELOPMENTS

Gene transfer in mice had been achieved by microinjecting a DNA containing solution directly into one of the two pronuclei of the zygote (for technical details see Hogan, *et al.*, 1986). In the mouse the optimal time for collecting pronuclear embryos suitable for microinjection is approximately 15 hours after fertilization. Following microinjection, mouse embryos are transferred immediately back into the oviduct of pseudopregnant females. Alternatively, microinjected mouse embryos can be cultured overnight to the two-cell stage and then transferred into a recipient.

Superovulation protocols were well worked out for the production and collection of sheep embryos around seven or eight days old, but protocols had to be developed for the collection of fertilized sheep eggs at the pronuclear stage. The techniques required for the transfer of day six to eight blastocysts to a recipient ewe's uterus were established, but again there was a need to develop the most optimal protocol for the transfer of day one embryos. Lastly, sheep zygotes had not previously been studied to determine if pronuclei were visible during the one-cell stage or, if they were visible, at what time were they optimum for microinjection. Thus, when work was started to produce the first transgenic sheep, there was considerable background information which needed to be established empirically before standardized protocols could be established.

The first transgenic sheep was reported in 1985 by Hammer, *et al.* This single animal demonstrated that it was at least possible to transfer a foreign gene, in this case a mouse metallothionein human growth hormone (mMThGH) fusion gene, into sheep albeit at a very low efficiency. To date there are reports in the literature from three groups that have successfully, and routinely, produced transgenic sheep resulting in sheep carrying nine different transgenes (Table 1). However, as reviewed by Rexroad and Pursel (1988) the efficiency of producing genetically engineered sheep is still low, with only about 0.75 to 1 percent of the microinjected eggs transferred into recipient ewes resulting in the birth of transgenic lambs. An even lower percentage of embryos transferred result in lambs expressing the incorporated transgene, as not all transgenic animals express the new gene.

To reach this point essentially required a better understanding of the timing of the early development of the sheep embryo, rather than new technological breakthroughs. A treatment step requiring the administration of gonadotropin releasing hormone (GnRH) to the superovulation protocol was added to ensure that ovulation in the donor ewes was more nearly synchronous, thus greatly increasing the number of fertilized eggs collected at the pronuclear stage (Nancarrow, *et al.*, 1984). Centrifugation

Table 1. EFFICIENCY OF PRODUCTION OF TRANSGENIC SHEEP

Transgene	Number of injected embryos transferred	Lambs born	Lambs transgenic*	Transgenic lambs expressing Number	%
mMT–hGH[a]	1032	73	1	—	—
mMT–bGH[a]	842	47	2	2	100
mMT–hGRF[a]	435	63	9	1/7	14
mMT–bGH[a]	247	42	11	3	27
mAL–hGRF[a]	171	16	4	2	50
mMT–Tk[b]	150	29	1	—	—
BLG–FIX[b]	307	52	4	2/2	100
BLG–alpha 1AT[b]	49	11	1	1	100
Mtla–oGH5[c]	1089	83	4	0	0
Mtla–oGH9[c]	409	23	3	3	100
Totals	3699	439 (11.8)	40 (1.1)	14	35

[a]Rexroad, *et al.*, 1990.
mMT = mouse metallotheionein 1
mTF = mouse transferrin
mAL = mouse albumin
hGH = human growth hormone
bGH = bovine growth hormone
hGRF = human GH releasing factor

[b]Simons, *et al.*, 1988
BLG = ovine beta–lactoglobulin
Tk = thymidine kinase
FIX = human blood clotting factor IX
alpha 1AT=human alpha 1–antitrypsin

[c]Murray, *et al.*, 1989
Mtla = ovine metallothionein 1a
oGH = ovine growth hormone
Mtla–oGH5 transgene incorporated SV40 viral enhancer sequences, while the Mtla–oGH9 construct did not.

* Nancarrow, *et al.*, (1991) have since produced 1 additional transgenic sheep with each of the Mtla–oGH constructs, with the Mtla–pGH9 individual expressing the transgene. However, the numbers of injected embryos transferred and number of lambs born were not reported.

of the zygotes was tried to improve the visualization of the pronuclei (Nancarrow, *et al.*, 1984), but in contrast to pig embryos (Wall, *et al., 1985)* this was not helpful. Suitable visualization of pronuclei in sheep zygotes can be obtained by the critical use of differential interference contrast optics, whereby over 90 percent of pronuclear stage eggs can be successfully microinjected (Simons, *et al.*, 1988; Murray, *et al.*, 1989; Rexroad, *et al.*, 1989).

Part of the inefficiency in the production of transgenic farm animals results from imperfect handling techniques. Immediate transfer of embryos derived from superovulation resulted in 47 percent continuing to develop compared to 88 percent of the embryos left *in situ* (Rexroad and Powell, 1991). Embryo survival after collection is no better when a complex medium is used (Medium 199 plus 0.10 percent fetal bovine serum) than for a simple phosphate buffered saline with serum medium. Microinjection further reduces viability (Rexroad and Wall, 1987; Walton, *et al.* 1987) resulting in approximately 12 percent of embryos surviving to lambing (Table 1).

Selection of embryos for transfer that were known to be viable and to have incorporated the transgene would reduce the costs associated with maintaining pools of recipients. Sheep embryos can be co-cultured for three days on oviductal cells with only about a 15 percent loss in viability if transferred to recipients that came into estrus 24 hours after the donors. Delayed recipients are necessary because co-culture somewhat retards embryonic development (Rexroad, *et al.*, 1990). However, as the early cleavage divisions are largely programmed by maternal gene products, this does not allow an adequate test of the embryos long-term viability. Longer periods of co-culture are possible, but only modest results with respect to long term development have been reported thus far (Gandolfi and Moor, 1987). If longer culture intervals of up to five to seven days can be reliably achieved, development in culture may become useful for predicting long-term viability.

A further benefit of a longer culture period would be the ability to remove a few cells from an embryo and use PCR technology to ascertain whether or not the microinjected DNA had been incorporated into the embryo's genome. Thus we would be in the position of transferring only those embryos to recipients that were known to be both viable and transgenic.

Although there are at present a number of inefficiencies in the technology used to produce transgenic livestock, they are likely to be overcome during this decade. However, the long-term application of this technology in animal agriculture is presently limited by our lack of knowledge about the genetic basis of rate-limiting steps that affect production traits. In most cases we do not know what genes to transfer in order to gain the maximum benefit to production efficiency.

GROWTH PROMOTING TRANSGENES TRANSFERRED INTO SHEEP

A number of fusion genes encoding growth promoting hormones have been transferred into sheep, including fusion genes encoding the human (Hammer, *et al.*, 1985), bovine (Rexroad, *et al.*, 1989) and ovine growth hormone (Ward, *et al.*, 1988; Murray, *et al.*, 1989) genes (GH) and the human growth hormone releasing factor (hGRF) (Rexroad, *et al.*, 1989). A number of these transgenes have used the mouse metallothionein I (mMT) promoter. Other promoters used include the control sequences from the sheep metallothionein 1a (Mt1a) gene (Ward, *et al.*, 1988) and the mouse albumin and transferrin genes (Rexroad, *et al.*, 1990).

These promoter elements were chosen in an attempt to direct the expression of the transgenes to specific organs, e.g. the liver, or to try to maintain a degree of external control over the level of expression of the transgenes. The mMT promoter had previously been shown to direct the expression of growth hormone to the liver, kidney and other organs in transgenic mice (Palmiter, *et al.* 1982, 1983). The promoter elements of the various metallothionein genes can be stimulated to increase the level of expression of a linked coding region by the addition of heavy metals to the diet or drinking water. The mMt promoter had the disadvantage of being leaky, that is there was always sufficient expression of the growth hormone transgenes in the absence of heavy metal stimulation to promote growth of the mice to approximately 1.5x the size of non-transgenic littermates (Palmiter, *et al.*, 1982, 1983). However, in the absence of direct experimental evidence, the pattern of expression of the various mMT–GH and GRF transgenes in sheep could not be predicted.

The sheep Mt1a promoter was selected by the Australian group for basically three reasons. First, ideally the expression of a transgene affecting growth should be under external control so that transgenic animals only express the transgene when it is economically advantageous to have them respond. Secondly, we thought that a more tightly controlled a transgene would be, the less likely to have adverse affects on the individuals carrying it. Thirdly, it was felt that transgenes composed entirely of sheep gene sequences would be more acceptable to lay persons, in particular consumers. For these reasons, we elected to use the sheep Mt1a promoter as it was known to have a lower basal level of expression than the mouse MT promoter, while still retaining its heavy metal inducibility (Peterson and Mercer, 1986).

The mouse albumin and transferrin promoters were selected to try to limit the expression of the transgenes to liver tissue in transgenic animals. These promoters were tried to attempt to limit the degree of undesirable side effects that had been observed in transgenic pigs carrying mMT based GH transgenes (Pursel, *et al.*, 1987).

THE CONSEQUENCES OF GROWTH PROMOTING TRANSGENES EXPRESSION IN SHEEP

Thirty-six lambs have been produced by microinjecting fusion genes encoding either growth hormone or growth hormone releasing factor (Table 1). Nine of these transgenic sheep carried transgenes that expressed high levels of growth hormone, while three lambs expressed growth hormone releasing factor. Many of the 24 non-expressing sheep have transmitted the transgene to their progeny, which also failed to express the transgene (Rexroad, *et al.*, 1989, 1990; Murray, *et al.*, 1989; Nancarrow, *et al.*, 1991).

The transgenic lambs expressing either GH or GRF grew at approximately the same rate as non-transgenic controls, even though they had circulating plasma GH levels from 3 to >1,000 times higher than found in control animals (Rexroad, *et al.*, 1989; Murray, *et al.*, 1989). The expressing transgenic animals also had elevated levels of circulating IGF–I and insulin (Rexroad, *et al.*, 1990; Nancarrow, *et al.*, 1991). Additionally, we have observed that plasma levels of prolactin and the thyroid hormones were depressed in expressing females, but were elevated and not different from controls in a single expressing transgenic male. Lower levels of thyroid hormone are normally associated with a reduced basal metabolic rate (BMR), but in the case of two of the GH expressing females, BMR was increased about 30 percent and metabolic heat production by 20–50 percent (Nancarrow, *et al.*, 1991).

The secretion of FSH and LH from the anterior pituitary gland was normal in both males and females, as were the serum levels of the sex steroids. However, both expressing males and females appear to have delayed sexual development when compared to controls (Nancarrow, *et al.*, 1991). This is perhaps similar to the situation observed in expressing growth hormone transgenic pigs where females remain anestrous and males lack libido, but produce viable spermatozoa (Pursel, *et al.*, 1990).

One of the goals of this research was to determine if GH transgenic sheep were more feed efficient and produced less fat than currently available animals, as has been observed in GH transgenic pigs (Pursel, *et al.*, 1989). Expressing transgenic sheep do not appear to have increased feed efficiency (Rexroad, *et al.*, 1989), but they do have significantly reduced amounts of body fat (Ward, *et al.*, 1990; Nancarrow, *et al.*, 1991).

The continuously high levels of circulating growth hormone observed in the expressing transgenic sheep has lead to severe health problems (Nancarrow, *et al.*, 1991). None of the twelve expressing animals has attained puberty, with all of them dying before one year of age. The cause of death has varied, but there are clear data that the over-expression of GH adversely affects liver, kidney and cardiac function (Nancarrow, *et al.*, 1991). Plasma levels of insulin and glucose are also abnormal, indicating an inability to maintain serum sugar levels that results in a diabetic condition (Rexroad, *et al.*, 1990).

Although the efficiency of producing transgenic sheep is low, it is apparent that foreign genes can be reliably transferred into sheep and expressed. The work with growth hormone transgenic sheep and pigs referred to here clearly shows that basic research is still required in a number of areas. The efficiency of the procedure needs to be improved to reduce the cost of producing genetically engineered animals.

More importantly, further work is required to identify promoter elements that, when used in transgene construction, will give a sufficient degree of control over the tissue specificity, developmental timing and level of expression of the transgene. Clearly, the uncontrolled over-production of genes encoding hormones will mostly likely be detrimental to the animal as observed in these studies.

Although there is probably no limit on the types of genes that can be transferred into livestock species, further research is required to identify the genetic basis of the rate-limiting steps affecting economically important traits. If genetic variation exists in a species that can positively affect the trait of interest, then selective breeding can be practiced. If variation does not exist and a gene from another species can be identified that will positively affect the trait, then gene transfer becomes the technique of choice. To date, there have only been a few candidate genes identified for transfer into livestock, such as the two genes of the cysteine biosynthetic pathway for wool production in sheep (Ward, *et al.*, 1990), that can reasonably be expected to significantly improve a production trait. The one exception may be for the dairy industry, where there are a number of potential ways genetic engineering techniques may be used to alter or improve milk (Wilmut, *et al.*, 1990).

FUTURE CONSIDERATIONS

The demonstration that genetic engineering techniques can be successfully applied to livestock animals raises a number of questions in such diverse areas as ethics and animal rights, economics and the future directions and applications of this research. As scientists involved in this work, we are satisfied that the information collected and the potential applications will be valuable to mankind. As our population increases and the resources of the Earth decrease, agricultural production must become more efficient. Does the non-scientist accept this point of view? Do we have the right to alter the genetics of our farm animals to increase production efficiency, or for that matter, will the public accept the agricultural use of transgenic animals?

Apart from these questions on the morality and ethics of modifying a species' genome, a number of other of questions remain to be answered. Who should pay for this research, which is clearly long-term and expensive, and who will benefit? At present, most of the significant work has been carried out in laboratories with a high level of direct government funding, for example the USDA/ARS laboratory at Beltsville, Maryland or the CSIRO's Division of Animal Production in Australia.

Transgenic experiments using livestock species requires long-term research projects, in part due to the reproductive cycles of the animals and in part because of the need to carry out a careful examination of the consequences of transgene expression in at least two or three generations of animals. Because the work is long-term with little or no prospect of early returns, only a few companies are currently prepared to contribute financially to this work. At a time when federal government granting agencies are experiencing a drop in the level of projects they can support, they are unwilling to commit support for the eight to ten years that these types of projects require.

It is not sufficient to conclude that the basic research should be done in less costly laboratory animals like the mouse and only scaled up to large animals after the transgene is proven. The Mt1a–oGH transgene acts as a completely controllable gene in the mouse (Shanahan, *et al.*, 1989), yet was totally uncontrollable in transgenic sheep (Murray, *et al.*, 1989). Furthermore, Mt1a–oGH transgenic mice, when switched on, grow to about 1.5X larger than controls, are more feed efficient with respect to lean production, deposit as much fat as controls (Pomp, *et al.*, submitted) and do not suffer any apparent health problems. Thus, each transgene construct intended for use in farm animals will ultimately need to be tried in farm animals during the research phase. It is only after the basic patterns of expression and consequences of a transgene's expression have been assessed in a livestock species can valid predictions of the worth of that transgene to agriculture be confidently made.

Thus, if this work is deemed important, who will pay for the research? Additionally, how will a suitable transgene be bred into the national herds? Who will control the flow of a valuable transgenic germplasm into the national herds? And how will this dissemination be paid for?

The production of transgenic animals also raises a number of legal or paralegal questions, both in the United States and in the international arena. For example, in the United States it is legal to patent a transgenic animal, but in Europe it is not. This raises a number of issues. How different must two lines of transgenic animals be for separate patents to be issued? How are patents on livestock going to be enforced in a loosely regulated industry such as farming? What are the consequences in terms of international trade?

The broad scientific questions are clear. We need to identify the rate limiting steps affecting production traits and appropriate genes that, when transferred, will overcome these limiting steps. In addition, we need to build up a library of tissue and developmentally specific promoter elements that can be used to drive the appropriate level of expression of a transgene. However, the societal answers to the questions we have raised here, and the setting of clear priorities and directions for the application of transgenic technology in animal agriculture have not yet been addressed.

REFERENCES

Costantini, F. and E. Lacy. 1981. Introduction of a rabbit beta globin gene into the mouse germ line. *Nature*, 294:92–94.

Gandolfi, F. and R. M. Moor. 1987. Stimulation of early embryonic-X-development in the sheep by co-culture with oviduct epithelial cells. *J. Reprod. Fert.* 81:23.

Gordon, J.W., G. A. Scangos, D.J. Plotkin, J. A. Barbosa and F. H. Ruddle. 1980. Gentic transformation of mouse embryos by microinjection of purified DNA. *Proc. Natl. Acad. Sci.* USA 77:7380–7384.

Hammer, R.E., V. G. Pursel, C. E. Rexroad Jr., R. J. Wall, D. J. Bolt, K. M. Ebert, R. D. Palmiter and R. L. Brinster. 1985. Production of transgenic rabbits, sheep and pigs by microinjection. *Nature*, 315:680–683.

Hogan, B., F. Costantini and E. Lacy. 1986. *Manipulating the mouse embryo: A laboratory manual.* Cold Spring Harbor Laboratory.

Murray, J.D., C. D. Nancarrow, J. T. Marshall, I. G. Hazelton and K. A. Ward. 1989. Production of transgenic merino sheep by microinjection of ovine metallothionein–ovine growth hormone fusion genes. *Reprod. Fert. Devel.* 1:147–155.

Nancarrow, C.D., J. D. Murray, M. P. Boland, R. Sutton and I. G. Hazelton. 1984. Effect of gonadotrophin releasing hormone in the production of single-cell embryos for pronuclear injection of foreign genes.in *Reproduction in Sheep.* D.R. Lindsay and D.T. Pearce eds. Aust. Acad. Sci., Canberra. pp. 286-288.

Nancarrow, C.D., J. D. Murray, M. P. Boland, I. G. Hazelton and R. Sutton. 1984. Towards gene transfer into ruminant embryos: Effect of centrifugation. *Theriogenology* 21: 248.

Nancarrow, C.D., J. T. A. Marshall, J. L. Clarkson, J. D. Murray, R. M. Millard, C. M. Shanahan, P.C. Wynn and K. A. Ward. 1991. Expression and physiology of performance regulating genes in transgenic sheep. *J. Reprod. Fert.*, Suppl.43:277–291.

R. D. Palmiter, R. L. Brinster, R. E. Hammer, M. E. Trumbauer, M. G. Rosenfeld, N. C. Birnberg, and R. M. Evans.1982. Dramatic growth of mice that develop from eggs microinjected with metallothionein–growth hormone fusion genes. *Nature*, 300:611–615.

Palmiter, R.D., G. Norstedt, R. E. Gelinas, R. E. Hammer, and R. L. Brinster, . 1983. Metallothionein–human GH fusion genes stimulate growth in mice. *Science*, 222:809–814.

Peterson, M.G. and J. F. B. Mercer. 1986. Structure and regulation of the sheep metallothionein–1a gene. *Eur. J. Biochem.* 160:579–585.

Pomp, D., C. D. Nancarrow, K. A. Ward and J. D. Murray. Growth, feed efficiency and body composition of transgenic mice expressing a sheep metallothionein 1a–sheep growth hormone fusion gene. Submitted manuscript *Livestock Production Science* (in press). 1992.

Pursel, V.G., C. E. Rexroad Jr, D. J. Bolt, K. F. Miller, R. J. Wall, R. E. Hammer, C. A. Pinkert, R. D. Palmiter andR. L. Brinster. 1987. Progress on gene transfer in farm animals. *Vet. Immunology Immunopathology*, 17:303–312.

Pursel, V.G., C. A. Pinkert, K. F. Miller, D. J. Bolt, R. G. Campbell, R. D. Palmiter, R. L. Brinster and R. E. Hammer. 1989.Genetic engineering of livestock. *Science* 244:1281–1288.

Pursel, V.G., R. E. Hammer, D. J. Bolt, R. D. Palmiter and R. L. Brinster. 1990. Integration, expression and germ-line transmission of growth-related genes in pigs. *J. Reprod. Fert.* Suppl. 41:77–87.

Rexroad, C.E. Jr. and A. M. Powell. 1991. Effect of serum-free co-culture and synchrony of recipients on development of cultured sheep embryos to fetuses. *J. Animal Sci.* 69:2066–2072.

Rexroad Jr, C.E. and V. G. Pursel. 1988. Status of gene transfer in domestic animals. *11th Inter. Congr. on Animal Reprod. and A.I.*, Dublin. 5:28–35.

Rexroad, C.E. Jr. and R. J. Wall. 1987. Development of one-cell fertilized sheep ova following microinjection into pronuclei. *Theriogenology* 27:611–619.

Rexroad, C.E. Jr., R. E. Hammer, D. J. Bolt, K. E. Mayo, L. A. Frohman, R. D. Palmiter and R. L. Brinster. 1989. Production of transgenic sheep with growth-regulating genes. *Mol. Reprod.Devel.* 1:164-169.

Rexroad, C.E. Jr., R. E. Hammer, R. R. Behringer, R. D. Palmiter and R L. Brinster. 1990. Insertion, expression andphysiology of growth-regulating genes in ruminants. *J. Reprod.Fert.* Suppl. 41:119–124.

Rexroad, C.E. Jr., A. M. Powell, R. Rohan and R. J. Wall. 1990. Evaluation of co-culture as a method for selecting viable microinjected sheep embryos for transfer. *Animal Biotech.* 1:1–10.

Shanahan, C.M., N. W. Rigby, J. D. Murray, J. T. Marshall, C. A. Townrow, C. D. Nancarrow and K. A. Ward. Regulation of expression of a sheep metallothionein 1a–sheep growth hormone fusion gene in transgenic mice. *Mol. Cell Biol.* 9:5473–5479.

Simons, J.P., I. Wilmut, A. J. Clark, A. L. Archibald,J. O. Bishop and R. Lathe. 1988. Gene transfer into sheep. *Bio/Technology* 6:179–183.

Wagner, E.F., T. A. Stewart and B. Mintz. 1981. The human B–globin gene and a functional viral thymidine kinase in developing mice. *Proc. Nat. Acad. Sci.* USA 78:5016–5020.

Wagner, T.E., P. C. Hoppe, J. D. Jollick, D. R. Scholl, R. L. Hodinka and J. B. Gault. 1981. Microinjection of the rabbit B–globin gene into zygotes and its subsequent expression in adult mice and their offspring. *Proc. Natl. Acad. Sci.* USA 78:6376–6380.

Wall, R.J., V. G. Pursel, R. E. Hammer and R. L. Brinster. 1985. Development of porcine ova that were centrifuged to permit visualization of pronuclei and nuclei. *Biol. Reprod.* 32:645–651.

Walton, J.R., J. D. Murray, J. T. Marshall and C. D. Nancarrow. 1987. Zygote viability in gene transfer experiments. *Biol.Reprod.* 37: 957-967.

Ward, K.A., C. D. Nancarrow, J. D. Murray, C. M. Shanahan,C. R. Byrne, N. W. Rigby, C. A. Townrow, Z. Leish, B. W. Wilson, N. M. Graham,P. C. Wynn, C. L. Hunt and P. A. Speck. 1990. The current status of genetic engineering in domestic animals. *J. Dairy Sci.* 73:2586–2592.

Ward, K.A., J. D. Murray, C. M. Shanahan, N. W. Rigby and C. D. Nancarrow. 1988. The creation of transgenic sheep for increased wool production in *The Biology of Wool and Hair* G.E. Rogers, P.J. Reis, K.A. Ward and R.M. Marshall eds. Chapman & Hall, London and New York. pp. 465-477.

Wilmut, I., A. L. Archibald, S. Harris, M. McClenaghan, J. P. Simons, C. B. A. Whitelaw and A. J. Clark. 1990. Modification of milk composition. *J. Reprod. Fert.* Suppl. 41:135–146.

ANIMAL GROWTH BIOTECHNOLOGY IN A QUANDARY
ISSUE DIMENSIONS & OPTIONS

ABSTRACT

Agriculture has enjoyed dramatic successes through biotechnology in recent decades and consumers have come to expect a bountiful, appealing, nutritious, healthful, economic, convenient and safe food supply. Consumers now demand products which are desirable in composition and value, and safe and wholesome.

In order to meet these demands, several types of animal growth biotechnologies have been safely employed in food production systems for several decades and technologies now in several stages of development involve new types of growth regulators and recombinant approaches. While the "new" biotechnologies have been positioned as somehow different from those currently in use, attention and concern in producer, consumer, political and activist arenas has arisen regarding both present and emerging growth regulating technologies, whether or not they involve recombinant technology. The European Economic Community (EEC) ban on anabolic growth regulators for economic and trade purposes under the guise of other issues (i.e., safety) and moratoriums on the use of somatotropins for meat and milk production are examples of concerns and actions which have targeted animal growth technologies. Perceptions and facts are widely divergent on these issues. Production of foods in systems untampered by humans surfaces as a common denominator in the discussions surrounding these issues.

To increase lean tissue and reduce fat deposition in animals, diet and health concerns along with animal efficiency in producing quality, safe, lean and healthful animal products all require immediate attention. The ability to produce highly palatable acceptable lean animal products is a

FLOYD M. BYERS

Animal Science
Texas A&M University
College Station, TX 77843

PAUL G. THOMPSON

Philosophy
Director, Center for Biotechnology Policy & Ethics
Texas A&M University
College Station, TX 77843

critical priority for the animal industry. Consumer calorie-consciousness requires a sincere effort on the part of the animal products industry to produce leaner animal products.

All technology implemented in the production system must be concurrently marketed to the final consumer as well as producers; this is seldom accomplished. We can no longer use technologies which are inconsistent with consumers' quality of life, and in the future, both the product as well as the systems used to produce it will need to be reflect consumers' needs and attitudes.

This paper explores the current status of the issues surrounding animal growth technologies and identify options and strategies through which these technologies may be successfully advanced in concert with the interests and perceptions of the diverse parties affected by these issues. Discussion papers on these issues were developed through a colloquia sponsored by the Texas A&M University Center for Biotechnology Policy and Ethics; these are excerpted below.

Some of the questions considered for discussion concerning animal growth regulating biotechnologies include:

—What are the implications in implementation or restriction of growth regulating technologies; will humans be better served?

—Why should these technologies be used? What are the benefits and who benefits?

—Who should be involved in the decisions and interchange involving use of these technologies for animals in food production? Should it involve those that do not use or consume foods produced with these products?

—How can all dimensions of the issues surrounding decisions on use of these technologies be communicated?

—What patterns and strategies might be useful in developing new approaches to deal with the development and implementation of growth regulation technologies?

INTRODUCTION

Understanding the issues surrounding animal growth biotechnologies requires a look back at the issues that have developed, how they developed over time, the basis for the issues as well as how and why they are important.

The target of animal industries is toward producing products that have a better consumer image and are aligned with consumer needs. For this discussion, the focus is on the use of biotechnology and other technologies to achieve these ends. The current aim isto produce what we call consumer-driven products—in composition and value—and systems which will be used to produce them. We are faced with the dilemma of choosing to use either traditional systems or biotechnology of some kind or a combination of both. Another question is, "*Can* we use them?" The concern of consumers is shifting from food itself to examining how it is produced.

Biotechnology might appear in food or it may be used to produce food. Consumers and other interested groups have certain expectations if we look at the growth area and animal products, most recently asking the question "Should biotechnology be used in food production?" and "Can we use biotechnology in a sustainable system or are we unable to develop an integrated system using biotechnology?"

The working group at Texas A&M studied some of the animal biotechnologies available and decided that certain technical issues, economic issues and ethical issues associated with the specific technology and those associated with change resulting from technology use would be addressed. Some very important political ramifications need to be addressed as well as food safety implications, diet and health implications and the issue of perception vs reality.

Contributors to this paper included Drs. Martin Terry, Jeff Burkhardt, Cathy Lemieux, Dave Hutchinson and Gordon Carstens. Most contributors gave seminars which raised questions concerning biotechnology use and suggested some of the possible implications of its use. They also asked who should be involved in making decisions about technology use and acceptability, and what benefits would be accrued from biotechnology use. Excerpts from the seminar papers are included to discuss our findings.

BACKGROUND PERSPECTIVES—GROWTH REGULATION BIOTECHNOLOGIES

Floyd M. Byers, Animal Science, Texas A&M University

Our food production systems were developed under the premise of scarcity. In developed countries where technology is used, innovation and technology have fostered an abundance of food and with it, a new set of societal values and expectations have emerged. The contemporary consciousness concerning biotechnology and resource use in food production systems worldwide reflects the dramatic success of our food production systems which provide an abundance of food with little involvement on the part of most consumers. This independence from the drudgery of searching for food provides much of society with the opportunity to pursue more fruitful endeavors toward advancement of humankind. A direct result of this dissociation is longer-term concern for the technology use in, and sustainability of, food production systems for future generations and maintaining and improving quality of life on the planet. Overriding issues concern application of biotechnology in food production, perceptions of resource use, and stewardship for resources planet-wide. Certainly, it was expected that different concerns would emerge when abundance rather than scarcity of food becomes the norm. Animal growth regulation biotechnologies have contributed to the abundance of desired animal products in our food supply.

As a consequence, the entire social contract between consumers and agriculture is now as never before in transition. The 1990s have become the decade of food safety and environmental awareness and consumers are demanding a safe, environmentally-sensitive and resource-conscious food supply and production system. Consumer-driven food products of desirable composition and value, that are assuredly safe and wholesome and that are produced in technologically, sociologically, and environmentally acceptable fashions are required. Consumers have an increasing interest in the diverse issues facing civilization, particularly those involving our food. How food is produced and the implications of biotechnology in food choices have surfaced as key issues about which the public now insists on being informed.

While science has responded with what is commonly described as an "information overload," little of this reaches consumers in easily understood ten second sound bites, resulting in an information vacuum on issues surrounding key aspects of food production. Since a vacuum will be filled, an ever expanding array of spokespersons representing key societal concerns and purportedly representing "consumers" are "carpet bombing" the communications media with targeted-simplified information (i.e., eat lower on the food chain, hormone-free, natural, etc.) to achieve specific and egalitarian objectives involving animal, cultural, environmental and ethical dimensions of food production. In todays' communication systems, perceptions gained through watching media messages using sounds, shapes and images take precedence over facts concerning food production issues, of which animal growth regulators are a component. In this forum, the credibility of the messenger has increasingly become a focus for discerning and forming perceptions and judgements on these issues. It would certainly be unfortunate if safe, efficacious technologies for producing safer and healthier consumer-desirable animal products were rejected by, or unavailable to, consumers on the basis of misinformation, disinformation and perceptions. Unfortunately, the value of these technologies, in use and in development, was not or has not been communicated to consumers with the same message penetration as the emotional appeal for "natural" food production systems, untampered by humans.

Current technologies used in animal (beef) production, for example (anabolic implants), modify (repartition) growth to allow production of leaner beef products with less fat. Similar technology does not yet exist in practice for other species. Emerging technologies, however, promise effective growth regulation options for beef, lamb, pork and poultry with possible applications for fish as well. Repartitioning of growth and consequent modification of animal products has received major attention in recent years, and clearly provides the most direct and efficacious mechanism to change the protein and fat content of animal tissues. The objective is to modify the patterns of growth in animals to produce less fat in animals and more lean animal products. While this is the eventual target of genetic engineering initiatives, systems employing these concepts (i.e., transgenic animals) are not likely to surface in the marketplace soon. A number of options are feasible in developing systems employing growth regulating biotechnology in several forms to produce leaner animal products, and include estrogens, zeranol, androgens (i.e., TBA), growth hormone, beta agonists, immunization and growth hormone releasing factor. All of these options have been investigated to varying degrees across animal species in developing targeted growth management systems to most efficiently produce desired leaner animal products.

The mechanisms involved in redirection of growth include modification of priorities for nutrient use for protein vs fat deposition, alteration of tissue turnover, modification of daily tissue deposition limits and modification of nutrient supply. Eventually, growth hormone, releasing factors for growth hormone, beta agonists and/or immunization strategies to remove negative feedback on growth (i.e., somatostatin) may provide additional mechanisms with which to regulate growth. They may work in concert with or replace current growth regulation technology, and these alternatives are currently in development.

Currently used estrogenic growth regulators (i.e., beef cattle), like growth hormone and beta agonists in development for several animal species, are effective repartitioning agents modifying growth by shifting nutrients from fat to protein accretion. Carcass animal products reflect accumulative growth from birth to slaughter. As a consequence, use of growth regulation biotechnologies from birth to slaughter provides lifetime growth regulation and provides the maximal redirection of nutrients from fat to protein and lean tissue production. The longer growth regulators are provided, the greater is the increase in total animal product lean with a simultaneous reduction in fat. Across several recent studies, the percentage of carcass fat was reduced by ~ 11 percent with current anabolic repartitioning implants (zeranol, estradiol 17ß). Concurrent with this reduction in carcass fat, the percentage of lean retail product was increased by 3.3 to 5.0 percent with these implant growth regulators. Reflecting the change in carcass composition and percentage retail product, the percentage of fat in the rib-eye muscle was reduced by 30 percent (from 3.8 to 2.6 percent). Trim fat (subcutaneous) and internal fat were similarly reduced.

The use of current growth regulators thus results in beef products with less trim fat (which consumers may eat or leave as plate waste) and with substantially less fat in trimmed lean muscle (i.e., rib steak) food products. Concurrent with this reduction in fat, cholesterol delivered to consumers will also be reduced, reflecting the two-fold higher cholesterol content of fat vs lean muscle. These growth regulation biotechnologies are therefore important components of integrated growth management systems to allow production of consumer-driven lean-diet health-desirable animal food products.

The growth regulators currently approved for use (for beef cattle) are either endogenous compounds already present in man and animals (estrogen, testosterone, progesterone) or are compounds developed through biotechnology to mimic these endogenous substances (zeranol, trenbolone acetate). These growth regulators are currently used in over 95 percent of all cattle on feed in the U.S. and in 50 to 75 percent of these same cattle during growth as calves and as stockers prior to feeding. None of these are ever fed to animals in the U.S.; they are instead placed in the ear, which does not normally enter the food chain. When used in cattle production, residues in meat are extremely low and less than naturally occurring levels in meat from cows and bulls. Levels of these substances (hormones) produced in people every day are many thousands to millions times greater than present in meat either naturally or as a result of use of a growth regulator in cattle. Also, other foods, especially vegetables, salad oil, etc. provide thousands of times more estrogen than meat from cattle whether receiving growth regulators or not, and less than 10 percent of what is consumed is absorbed by humans, so the contribution from beef is truly negligible.

The EEC imposed a ban on import of beef from the U.S. and other countries using anabolic growth regulators commonly referred to as "hormones." While the ban was originally launched under the guise of a "safety" issue the directive for the ban has been adopted by the EEC although all safety issues were dismissed long ago by both the EEC's own commission "The Lamming Commission" and by the U.S. governments regulatory agencies (the Food Safety and Inspection Service (FSIS) branch of the Department of Agriculture (USDA) and by the Food and Drug Administration (FDA)).

In contrast to the U.S., where biotechnology is tightly and efficiently regulated such that no violative residues were found in the past four years of USDA-FSIS' National Residue Program, a fraction of meat produced in the EEC contains unacceptable residues of compounds never cleared for use, some of which are known carcinogens. A safety issue exists with EEC animal products because of unapproved "cocktails" of many potent drugs directly injected into the muscle of growing animals on EEC farms, as a result of the ban on use of approved products instated during the past several years.

Recent data were summarized to assess the impact on the U.S. industry. In a summary of growth regulation studies at Texas A&M, the change in net return on a lean retail product basis averaged $96.68 per animal. This represents a net value to the U.S. beef products industry of approximately $2.5 billion. These data are consistent with results of a recent USDA study indicating a $2.4 to $4.1 billion reduction in net return on a retail product basis if currently approved growth regulators were not used in the U.S., depending on feeding and marketing management alternatives. Worldwide implications would obviously be much greater and this is borne out in the USDA study.

Clearly, when safe, approved efficacious biotechnology is banned to serve popular, protectionist or political purposes, only unapproved technology will be available for use. Use of approved safe growth regulators allows application of biotechnology to produce leaner animal products consistent with diet health needs of consumers. The ban on this technology in the EEC has resulted in the delivery of fatter beef products to European consumers, a situation inconsistent with the needs of U.S. (and other) consumers. Similar restrictions are forthcoming or are currently in place regarding the use of growth hormone based technology currently in development to modify meat animal products (i.e., EEC) or quantity of milk produced per animal (Minnesota, Wisconsin).

In producing environmentally sensitive animal products, the adoption of technology to reduce methane directly or growth regulators to enhance lean tissue growth reduce feed resources per unit of animal product and reduce the methane per unit of beef produced. Eliminating these technologies (i.e., growth regulation ban-EEC) results in a decrease in rates of lean tissue growth, more feed resources used and more methane per unit of beef or milk produced. Disallowing efficient meat production technology (i.e., growth regulators ban-EEC) or as suggested for milk or meat production (i.e., BST ban-EEC) would have unwanted resource (feed, energy, water, etc.) and environmental implications.

Excerpted From:

BAN ON GROWTH PROMOTANTS IN THE EEC—
THE ANATOMY OF A TECHNICAL TRADE BARRIER

Martin K. Terry, Animal Health Institute

INTRODUCTION

In December 1985 the Council of Ministers of the European Economic Community (EEC) enacted legislation in the form of a Directive (85/649/ EEC) banning the use of anabolic hormones in livestock production as of 1 January 1988. This Directive was misguided in its conception; by indiscriminately prohibiting all products, it in fact effectively eliminates safe, legitimate products. It thereby encourages the use of unapproved, potentially dangerous products, with the result that the health of consumers is placed in jeopardy — an ironic state of affairs to emerge from legislation ostensibly promulgated to allay "consumer anxieties". The hormone ban is inherently unenforceable (thus in essence inviting black market violations while economically penalizing those who comply with the directive), and is generally disastrous in its implications for European agriculture, world trade and the future of the technology upon which agricultural productivity depends.

FALSE PERCEPTIONS

Prior to 1980 the European public was largely unaware of the use of hormones in animal production. Public attention was drawn to the "estrogen scandal" in Italy in 1980, when DES residues were discovered in a batch of manufactured baby food and those residues were assumed to have originated from the use of DES as an anabolic agent in veal calves and to have been the cause of several cases of abnormal sexual development in school children. The Italian press fabricated a "link" between the use of hormones in livestock production and a hazard to the public health. The fact that this "link" was purely hypothetical — and that the hypothesis was not even compatible with the facts of the case — was soon forgotten, but the negative image of hormones remained in the mind of the public.

European consumer organizations called for an immediate ban on the use of all hormones as a high-priority item on their agenda. They neglected to note that the Italian estrogen scandal occurred in a country where a hormone ban was already established (and had been for nearly 20 years). The effectiveness of banning something that was already banned did not strike them as questionable.

SCIENTIFIC DELIBERATIONS AND POLITICAL AGENDAS

British Minister of Agriculture, Peter Walker, managed to work out an agreement for interim legislation (81/602/EEC) which did ban the stilbenes (such as DES) and thyrostatic agents, substances already banned in practically every country in the world, and required that the EC Commission conduct a detailed examination as to the "harmless or harmful effects" of the five compounds (estradiol-17ß, progesterone, testosterone, trenbolone acetate and zeranol).

An important feature of the 1981 Directive was that it upheld scientific criteria as the proper basis for the regulation of animal production drugs such as hormones. A scientific committee chaired by Professor Eric Lamming of the University of Nottingham began work in 1981 and produced an interim report in September 1982. This report found the endogenous hormones (estradiol-17ß, progesterone and testosterone) to present no hazard and requested further data on trenbolone acetate and zeranol.

The information requested by the "Lamming Committee" was supplied to the EC Commission by January 1984 and the Lamming committee was reconvened by the Commission to complete their report on trenbolone and zeranol. However, the Commission issued a proposal for new legislation (COM(84)295) without waiting for the Lamming Committee's final report. The Commission proposed a directive that would have authorized the use of the endogenous hormones (on the basis of the first Lamming Committee report) and would have banned the xenobiotics, on the basis of the 1982 Lamming report.

The Lamming report was not finished by the autumn of 1985, and the European Parliament voted an opinion which called for a ban on the use of *all* hormones, including those previously approved by the Lamming Committee.

The Parliamentary vote calling for a total ban took place on October 11, 1985. On October 30, the Commission issued a revised proposal (COM (85) 607) which similarly called for a total ban.

Thus the Commission completely eliminated the role of science from the decision-making process on hormones. The new proposal implicitly rejected the scientific evidence it had *previously accepted* as the proper basis for definitive legislation on hormones. The new proposal, though dealing with a matter of central importance to the protection of the public health, now seemed to be based primarily on political and economic considerations.

WHENCE THE BEEF MOUNTAIN?

"Intervention stocks" consist of subsidized meat which is bought and stored by the EEC, at taxpayers' expense, until it is so old as to have lost most of its value, at which time it is disposed of to a third country at a price far below the world market price for fresh beef.

By the latter half of 1985, when the European Parliament was debating the hormone issue, the amount of beef in intervention storage had reached a record *700,000 metric tons.* This "beef mountain" was a major source of embarrassment to European politicians, who realized what the problem of overproduction of beef was due to the Common Agricultural Policy (CAP) system of beef intervention payments which directly rewarded overproduction.

The consumer lobbyists offered the Europoliticians a solution to their dilemma. The idea was to ban hormones, which they said, were dangerous in the first place, and which also contributed to the overproduction reflected in the beef mountain.

The Europoliticians did *not* find the linking of hormone use and the beef mountain to be simplistic. It was that perception, together with continuing doubts as to the safety of hormones (in the absence of a definitive report from the Lamming Committee), that sealed the political fate of hormones in the EEC in 1985.

The result, legal or otherwise, was that on 31 December 1985 the Council of Ministers notified the EEC Member States that the directive banning the use of hormone implants in livestock as of 1 January 1988, was law.

All Member States were under an obligation to implement the Directive into their legal systems by 31 December 1987. All trading partners were under an obligation to ensure that any meat they exported to the EC after 1 January 1988 would have to be certified as not having come from animals treated with hormones - a deadline later extended by one year to 1 January 1989.

In January 1987, the Standing Committee on Hormone Toxicology of the West German Society of Endocrinology met in Mainz to evaluate the new data on trenbolone and zeranol. Their published opinion was that these two hormones were suitable for use in the production of animals for human food.

The World Health Organization (WHO) and Food and Agriculture Organization (FAO) Joint Expert Committee on Food Additives (JECFA) met in Rome in June 1987 to evaluate the safety data on all five hormones. They concluded that the five hormones posed no danger to human health when proper practices were followed. These JECFA findings were published informally in December 1987, and the full JEFCA reports were officially published by WHO and FAO in March 1988.

At the World Veterinary Congress in Montreal on 18 August 1987, members of the Lamming Committee, acting in their private capacities, presented a summary of their findings on all five hormones for the first time. The full findings were then published in the journal *The Veterinary Record* on 24 October 1987.

All five hormones had been declared to be safe for use as growth promoters in cattle by the very scientists who had been selected by the European Commission to review them. What they were concerned about, as they indicated in their Interim Report in 1982, was that if safe hormones were banned, they would be replaced by unsafe products distributed through the Black Market.

Lamming's poignant prophecy has been abundantly fulfilled.In February 1987, 10 months before the deadline by which importers of beef into the EC would have to certify that their meat was not hormone-treated, the United States began consultations under the Standards Code of the General Agreements on Tariffs and Trade (GATT). The U.S. alleged that since the hormone ban was without any scientific basis, it amounted to a technical trade barrier because non-EC countries had to guarantee that their meat exported to the EC had not come from hormone-treated animals - which could not be verified by any objective test.

On 18 November 1987, acting on a proposal from the Commission, the EC Council of Ministers adopted a Decision to extend for 12 months the deadline by which exporters of beef products to the community had to comply with the hormone ban directive. The new deadline was midnight on 31 December 1988.

Exactly a week after the EC postponement was announced, the U.S. pursuant to an Order signed by President Reagan under section 301 of the amended Trade Act of 1974, published in the *Federal Register* a list of retaliatory tariffs which would apply to selected products from the EC.

Then the unpredictable happened. The United Kingdom of Great Britain (U.K.) had taken the Council to the European Court in March 1986, claiming that the hormone ban directive was invalid for a collection of alleged reasons. The European Court agreed with on of those reasons, a procedural one, and the Directive was invalidated.

At its next regular meeting, on 7 March 1988, the Council legislated by stampede.

On 16 December 1988, Dr. Lester Crawford, Administrator of the Food Safety and Inspection Service (FSIS) of USDA testified as a witness before the European Parliament's Committee of Enquiry investigating the hormone problem. Said Crawford in a press statement issued immediately after delivering his testimony: "If the European Community does not modify its hormone policy soon, meat trade between Europe and the United States will end..." He cited the re-appearance of DES in Europe as further evidence of public health problems caused by the hormone ban, and went on: "Ironically, the hormone ban itself has become the greatest threat to European meat consumers...Each year we must certify to the U.S. Congress that residue controls in exporting countries meet U.S. requirements. In the context or persistent reports of illegal use in EC countries of compounds with increasingly serious public health consequences, I don't think we can make that certification."

The U.S. had now taken two key steps. First, they had decided to retaliate by increasing tariffs on a range of European exports if American hormone-produced meat could not gain entry to the EC. Second, and separate from the retaliation, they were seriously considering banning further imports of European beef and veal owing to its probable contamination by dangerous Black Market hormones.

The amount of money involved is small in the context of total transatlantic trade worth $166 billion each year. U.S. beef product exports to the

EC amount to about $155 million. Much of this is offals—"speciality meats" such as kidneys, livers and ox tongues, and about a third of that is imported for use in European pet foods.

At its last meeting before the 1988 Christmas break, on 19 December 1988, the EC's Council of Ministers agreed to allow the U.S. to continue to import hormone-produced beef for use in pet food.

The United States' retaliation list, published in November 1987, imposed 100 percent tariffs on a wide range of European products to a value equivalent to the value of the U.S. meat refused entry into the EC. After the EC conceded over pet food, the U.S. correspondingly scaled down its retaliation to $97 million worth of tariffs. They targeted canned tomatoes, wine coolers, instant coffee, fruit juices, pet food, beef, tomato sauce and pork hams. The penalties were selected to fall mainly on Italy and West Germany, the two countries most passionately in support of the hormone ban. Between them, Italy and West Germany bore 70 percent of the retaliatory tariffs.

On 5 January 1989 USDA wrote to the Heads of the Veterinary Services of all European Community countries except those in Greece, Portugal and Luxembourg, saying that countries exporting meat products to the U.S. "must apply public health controls at least equal to those administered by the USDA." The letter pointed out that a proper residue control program is characterized by, among other things, an approval system for veterinary products based upon "rigorous scientific process". USDA pointed out that the Black Market reports in Europe... "...have indicated that EEC countries are not effectively controlling use of such compounds. Use of unapproved compounds is occurring across species and is now involving compounds of serious public health consequences."

The trade war over meat from hormone-treated animals was now separating itself from the dispute over the food safety consequences of the EC hormone ban. The U.S. Congress ranks domestic food safety higher than foreign trade. Dr. Crawford's concern over imports of European meat products contaminated by residues of Black Market substances since the EC hormone ban, is not a pawn on the chessboard of international trade.

Excerpted from:

ETHICAL ISSUES ASSOCIATED WITH BST

Paul G. Thompson, Center for Biotechnology Policy and Ethics, Texas A&M University

The controversy over Bovine Somatotropin (BST) involves disputes about many technical issues: Does milk from cattle treated with BST differ from milk now being produced at dairies across the nation? How quickly will dairymen adopt BST, and how will it affect economics of scale in dairy production? Will dairy production shift from traditional dairy production? Will dairy production shift from traditional dairy states to new locations? How will rural communities be affected? Are we sure that milk produced by BST-treated cows will be properly metabolized by human consumers? The answers to these and other technical questions are important because they bear upon questions of responsibility, social justice and human (and animal) well-being. There are some applications of these concepts on which our society enjoys a firm consensus, but other applications are notoriously contentious.

One fact of post modern society is that decisions by a few individuals to develop and disseminate new technologies can have enormous impact upon society as a whole. Although there are many instances where these impacts are predominantly beneficial, there are few (if any) occasions on which they are universally so. Decisions made far from the rural heartland, in corporate offices or in research facilities, can effectively determine that some producers will have to leave farming, that consumers will be buying new food products, and that rural residents, wildlife, and, indeed, society as a whole (including future generations) will have to cope with pollution or resource depletion. The development of new technologies produces new benefits, but it also causes unwanted consequences.

The BST case raises questions about three kinds of unwanted consequences. The first group of impacts are felt by producers who may be forced to adopt BST (or to cease production) because of competitive pressures. The second includes consequences for non-human animals. The third includes environmental impacts that may bear upon a large number of people, extending into future generations. Food safety issues have less to do with unwanted impact than with uncertainty.

Each kind of unwanted consequence is ethically controversial. In most sectors of the economy, producers would not expect to be shielded from the economic consequences of technical change. Farmers are raising a concern more typically voiced by organized labor, as when plant closings or new production lines lead to layoffs. Extension of ethical concern to farm animals and to environmental impacts are also hotly debated topics. Given the assumption that animal husbandry practices carried out for the purpose of human food production are generally acceptable, opponents of BST must show why this particular technology is cruel or alternatively, why traditional standards for animal care should be revised.

Since each of the three points where BST has been linked to unwanted outcomes is controversial, it will be useful to look at two ways of framing the ethical issues of responsibility.

The Intentional Action Model—Each of three types of unwanted consequence noted above involves impact upon individuals or groups who are powerless to avoid being affected. This is clearly the case with respect to farm animals and unborn generations of human beings, and it is true to a more qualified extent for small scale dairy producers, too.

Bovine somatotropin exists today because a few hundred individuals made research and development decisions over a half decade. The decisions and the actions that followed them were undertaken *intentionally*.

The individuals and groups that carried out research and development of BST are capable of actions that impose unwanted consequences upon others. The question is whether their possession of this capacity gives them an unfair or unjust form of power over these others. Examining BST, we find that the companies developing BST have far more economic power than do small dairy farmers.

Further, many of the scientists who have participated in the development of BST can be thought of as agents for the general public, at least, and perhaps for the farm community, in particular. Land-grant universities, where much of the BST work has been done, have historically accepted a further mandate to do science that will strengthen the development of rural communities. As such, dairy farmers may have a special claim upon these institutions. Although no one has argued that scientists have a special responsibility to look out for animals, it is not uncommon or unreasonable to think that the scientific community is well placed to look out for the general public's interests in environmental quality.

The fact that BST emerges as a technology for which these considerations are relevant *does not* settle the issue in favor of BST's critics. At most they establish a burden of proof in favor of the farm, animals and environmental interests that bear the costs of unwanted technical change. One might interpret the political debate that has raged over BST as a working out of just such an exchange of views in the democratic political process.

The Consequence Evaluation Model—The idea that any technical change produces winners and losers invites us to think of any new technology as a social bargain in which there are both costs and benefits. The key to evaluating this social bargain lies in identifying and measuring the full range of costs and benefits. While one should not underestimate the difficulty of making these judgements, the idea that a technology's costs and benefits can be compared with the costs and benefits of no technology provides an attractive way of discharging the imperative of responsibility for technical change.

When applied to BST, the consequence evaluation model would regard adverse, unwanted outcomes as costs that should be weighed against the projected benefits derived from lower milk costs. When one adopts a consequence evaluation model for assessing new technologies, the question of whether BST is an ethically acceptable technology hangs upon the answers to these technical questions.

When one compares total outcomes from two or more options (at a minimum, the options include BST and no BST) there is no obvious reason why intentional action should enter the picture at all. There are costs and benefits associated with the *status quo.* The possibility of taking very different approaches to the problem of unwanted outcomes earn itself feed policy controversy. It is far easier for two who have different interpretations of responsibility to talk past each other than it is for them to communicate.

ETHICS AND UNCERTAINTY

By its very nature, technical change involves unprecedented events. The reality of disagreement among alleged experts creates a situation in which a member of the lay public, lacking even the evidence to make informed judgements about who to believe, quite reasonably comes to regard all claims about the likely consequences of technical change with justifiable skepticism.

The unfortunate upshot is that political decisions about technology often become dominated by uncertainty. Technical uncertainty creates an opportunity for experts to disagree. When experts disagree, non-experts are faced with uncertainty about who to believe.

The ethical character of the BST debate changed drastically when claims about the safety of consuming milk from cows treated with BST became contested. Prior to the time that doubts about the safety of milk were raised, the issue was one of how to resolve issues of responsibility for the unwanted consequences of introducing BST. With the advent of controversy over food safety, the potential spectrum of affected parties increased dramatically.

What is even more important is the way that the ethical issue shifted from being one of dealing with unwanted consequences to one of uncertainty. There has never been serious scientific evidence to suggest that there would be unwanted health consequences for consumers of BST milk. Consumer groups reacting to the food safety issue were not reacting to a health risk *per se*. Consumer groups were reacting to uncertainty, to a problem in deciding who to believe about BST and milk.

In understanding the way that ethics bear upon risk and uncertainty, it is crucial to see that the consumer's information about the safety of BST is all subject to a conditional probability that the source of that information is either ignorant or, worse, willing to deceive them. Many technical authors have taken to describing the difference between risks calculated on the basis of scientific evidence and risks calculated on the basis of corrigibility of human beings who report scientific findings as a distinction between "real" and "perceived" risk. This choice of words is sometimes unfortunate, for it can be taken to imply that the lay person is responding to extraneous and irrelevant evidence.

Given the background of the uncertainty problem faced by food consumers and consumer advocates, it is not surprising that the issue evolved into a debate about the risks of BST and milk. The scientific community has come to view risk issues as an expected value problem, and this is the way that the food safety issue for BST has been approached. While there are clearly many cases in which the assessment of expected values is the right approach to take for food safety, neglect of alternatives, burden-of-proof approaches may have been a better choice for BST.

Responding to uncertainty problems with technical risk assessments is, to a person unschooled in probability and consequence evaluation, little more than saying, "Trust me." Uncertainty issues are politically fractious and intense. It is far from clear that burden of proof approaches would have fared better. It is possible, however, that an agreement to label BST milk might have been interpreted as a gesture of good faith, one that empowers consumers to judge the risks of BST and milk for themselves. While labels satisfy a burden of proof for acceptable risk, requiring labels may have policy implications that are themselves unacceptable.

Democratic political theory has evolved around the concept of a social contract. BST has tested that social contract. Researchers and private companies have undertaken research and development on BST with the expectation that, if the product finds market acceptance, their efforts will be rewarded. While it is reasonable that they should have expected to deal with some of the unwanted consequences of BST, it was not reasonable to expect that food safety issues would be among them. The emergence of uncertainty and, in turn, the food safety issue is evidence of trouble in the contract. It is evidence of a lack of confidence in science and in science institutions. This is a development that should be viewed as quite serious, not only for science, but for the foundations of democratic institutions. The problem is that both commerce and political decision-making require a certain amount of trust. Whatever the causes, and however just or unjust the suspicion of science might be, the largest and most serious ethical issue associated with BST is the matter of trust. All the other ethical questions feed into this one.

One way of solving this problem is to build a high wall between that component of science which is in a position of public trust, and that portion of science which is involved in the development of technologies that may produce unwanted consequences. Public science, conducted at nonprofit institutions, would enjoy public confidence. Private science, conducted in the private sector, would be held to the same degree of accountability normally expected of commercial activity. The flaws in this solution are complicated and subtle. Features of contemporary science that make this ideal very difficult to achieve include:

1–Scientific research does not respect the public/private divide.

2–Enforcing a strong separation between public and private science is impractical.

3–Public science institutions are finding it necessary to cultivate private sources of research funding.
4–A strong separation between public and private science sectors might well weaken public science.

Therefore the dilemma is deep. The tension between the regulatory and the technology stimulating roles of science erodes public trust in science institutions. At the same time, any solution to this problem must be sensitive to the delicate network of personal relationships that makes science possible.

CONCLUSION

The ethical controversy over BST arose because, like many technologies, it may produce some effects that are unwanted. There is no reason to think that the unwanted consequences of BST are particularly dramatic or extreme, but the fact that decision makers within public research organizations or private companies can affect others makes these unwanted outcomes an issue of some significance. The importance is increased, however, because of the food safety questions that have been raised, and because of the climate of uncertainty that they generated. It is the uncertainty issue that truly threatens to keep BST off the market at this writing, and which the developers of the technology had no reason to expect. This, in turn leads to the question of trust that is crucial to democratic institutions. This is not to say that the success or failure of U.S. constitutional democracy hangs upon the BST decision, but it is to say that this policy problem is an example of a problem that can be expected to recur in the future.

Excerpted from:

ETHICS OF TECHNICAL CHANGE: *THE CASE OF BST*

Jeff Burkhardt, Food and Resource Economics, University of Florida

Recombinant Bovine Somatotropin (BST) has been the subject of controversy for over five years. Most if not all of the potential scientific, political, economic, and ethical risks and benefits associated with widespread use of the compound are well-documented. Yet, as is evidenced by the continuing stream of professional journal and newspaper articles, and workshops and symposia around the world, the issues associated with BST remain topics of heated debate. There are several reasons for this. That BST is one of the first commercially-feasible products from biotechnology is certainly part of the case. There are, however, many more important dimensions to "The Case of BST" than its biotechnology connection.

One important condition in this regard is the larger society's "philosophy of technology" which serves to guide both individual and institutional decisions. One such philosophy - to be referred to as "productionism"—has come to dominate decisions about both agriculture and agricultural research. Although there are strong historical as well as ethical reasons behind the productionist idea, this view has come under increasing philosophical (as well as political) attack. It is at this philosophical level that the major ethical concerns pertaining to the case to BST are raised. For, in the face of ethical challenges to productionism, we, as individuals and as a society, must again decide whether "more technology" and "more products" are to remain social and individual goals. For many social analysts, and in many governmental accounts of future possibilities for our society's technological mode, there seems to be an assumption that particular kinds of changes are inevitable. The focus on "trends," "tracks" and "outcomes" suggests that particular kinds of change *will* occur.

As important as the notion of choice is in the analysis of technical change, it is a mistake to think that all particular technical changes are the result of planning or conscious design. Technical changes occur for a variety of reasons, motives and beliefs, not all of which are calculated, planned and informed. The notion of a "technology treadmill" is understandable in this light. In effect, because of our prior decisions and the decisions of others, we must choose technical change. As the stream of new technologies continues, we will continue to have to make those choices.

At the individual level, "the ethics of technique" pertains to the actual practice of using a tool or technique. The ethics associated with change in technique at this level are principles governing learning about and monitoring the use of the new tool or strategy, so that bad consequences are avoided.

"Technical ethics" also covers the decisions groups of individuals make to adopt techniques in the first place: is the choice of a particular "new" technique or tool justifiable from the point of view of the array of ethical values which surface in an individual's interactions and connections with others in a larger community?

A third dimension of the ethics of technique involves the decisions groups of individuals make to allow or legitimize the adoption of new techniques on a larger societal scale. Decisions of this sort are made by "invisible hands" or "visible hands." For a variety of ethical reasons (including freedom of choice), markets are judged to produce the "best results": markets deliver what consumers want, for whatever reasons they want, including ethical (and unethical) reasons.

Markets only deliver or governments only authorize techniques or products which have already been at least conceptualized. The fourth dimension of "technical ethics" thus concerns the decisions individuals or groups make with respect to research on and development of new techniques and alternative ways of doing things. The judgement that a particular technique or tool should be developed in the first place is certainly made in the hope or expectation that it will be "induced" into a political-economic system once developed. "Economic" interests in the success or profitability of the new technique undoubtedly predominate. The value-context (e.g., "mission") of research institutions themselves is also relevant, as is whether or not the decisions are made in private, or in the public arena. This reflects consideration of ethical values subscribed to by the larger society: toward what techniques are potential consumers and the larger public disposed?

The last and broadest dimension of the "ethics of technical change" pertains to the larger society's beliefs and dispositions toward new techniques. What dictates our willingness to accept, promote, certify, adopt and use new tools, technologies, or techniques? What principles explicitly or implicitly lead us to ratify particular kinds of technical changes? A basic question is: upon what principles and values do we judge even our basic philosophy of technology to be right or good?

The case of BST is a good "ethics case" with regard to technical change, for it includes all of the dimensions of individual and group decision-making, institutional legitimacy, and fundamental philosophy, although the last may be less than transparent at first sight.

The case of BST is standardly invoked in discussions of the ethics of the "new biotechnology", and genetic engineering in particular. This can be misleading. While BST is a product of procedures which involve genetic engineering, most of the issues associated with the case of BST are independent of its biotechnology connection. Of course, if there are major ethical problems associated with biotechnology *per se*, recombinant BST is not significantly different from "natural" BST harvested from the pituitary glands of cows. Unless we ethically reject all such "bioproducts," BST remains but a "new technology" in the broadest sense of that term. Ethical principles apply to the case of BST just as they apply to other new technologies, e.g., computer software.

The "ethics of technique" is raised initially in connection with the question of whether there is a right way and a wrong way to use BST. Analysts of BST have repeatedly noted that a significant change in "management" is associated with BST use: individualized dosages, careful monitoring, etc. This is significant from the point of view of economics and also from the point of view of ethics: BST carries with it some "rules."

The major ethical question associated with individual choice about BST is, however, whether there really is a choice. It may be that the economic and social situation in which farmers find themselves unfairly limit their opportunities for choosing alternative techniques. However, it would appear that farmers do have a choice not to use BST. That they choose to do so suggest that they have accepted the conditions historically prior to and philosophically behind the introduction of BST.

Most of the literature pertaining to the case of BST has focused on the "social ethics" features of BST legitimation and creation. The question of the "marketing of BST" and the responsibility of federal government agencies (USDA, EPA, FDA), state governments (e.g., Wisconsin, Minnesota) and the courts in deciding the legitimacy of the diffusion of this technique is of course a major element in this story. The conflict-ridden interactions of people, as well as the economic and ethical values undergirding the conflicts pertain to two basic questions: Why BST? Why *not* BST?

In the political sphere and in the larger public forum, the "case against BST" has been pressed by coalitions of dairy farmers, animal rights activists, the Foundation on Economic Trends, and most recently, the Consumer's Union. The arguments range from those highlighting the likely negative socio-economic effects on classes of dairy farmers, to indictments of the potential negative human and animal health effects, to larger questions about society's "permitting" biotechnologically-generated products of uncertain social, economic, and environmental worth. Across the stage, the National Institutes of Health, the developers of BST, and some producers' organizations, argue the domestic and international economic validity as well as human and animal safety of the compound. One senses at times that the future of farming, governmental legitimacy, scientific freedom and progress and the survival of corporate America are all on the line. Interestingly, it has been stated that had the creators of BST known how it would be received and handled by governments and the creators themselves, an alternative such as porcine somatotropin (PST) would have been promoted in the first place. However, as some social scientists and ethicists have noted, it is not simply a strategic error in not considering the potentially controversial consequences of research and product-development, it is an ethical one as well, in failure to have considered some farmers', animal activists, and the larger consuming public's perceptions and values.

Serving agriculture through productionism, rationalizing agriculture through "big science" and emphasizing efficiency (under a narrow economic definition of terms) have remained the reigning mission and strategy of public agricultural research, as well as the dominant direction of private, corporate research. American consumers have benefitted by having the lowest food-to-income ratio in the world. The technologies that have been developed, from biotechnologically-engineered plant varieties to agrichemicals to computer-driven irrigation systems and sophisticated post-harvest transport, wholesaling and retailing systems –and now BST –are heralded as having allowed agriculturalists to all-but-overcome the inherent risks involved in farming. There is little doubt that the spread of these research products into less-developed nations, and the ensuing "Green Revolution" which transformed not only agriculture but whole societies, have produced great benefits for many people.

In our case study, a critique of productionism underlies the challenge to BST. It is as well a critique of technophilia, and perhaps to the utilitarian ethical philosophy which provides it the best justification. The challenge is far from univocal or even articulate in the actual public arena, coming as it does from an eclectic collection of environmentalists, family-farm advocates, the Foundation on Economic Trends, and the like. Nevertheless, it is there: Why *not* BST? The intriguing thing about the "case of BST" is how quickly and deeply the ethical question of "how to use" BST leads to "why new techniques," ultimately to, what kind of people are we and what kind of society do we want?

Over and above agriculture and agricultural research, many features of "modern" society continue to be challenged from a diverse array of thinkers. Productionism, and productionist-oriented technical change have done wonderful things, but at great individual and societal expense. More is no longer better, if it ever were.

The anti-productionist critique may overstate the case. Moreover, in the context of an appraisal of "the case of BST" the issue may seem grossly out of place. BST is, after all, only a single product, affecting only a single, small sector of a significant but hardly history-moving industry.

Excerpted from:

RECOMBINANT BOVINE SOMATOTROPIN FOR USE IN BEEF CATTLE

Gordon E. Carstens, Animal Science, Texas A&M University

Advances in technology and continual commercial application of this technology to animal agriculture has enabled producers to deliver large quantities of animal food products to the consuming public. The steady supply of these nutritious animal products has contributed significantly to enhancing the quality of life. Whereas in the early 1900's many of the health problems were related to nutritional deficiencies, today many of the health problems are associated with the over-consumption of calories, saturated fatty acids and cholesterol.

Whereas research in the past has focused primarily on methods to maximize productivity of food-producing animals to lower cost of production for the producer and cost of food products for the consumer, future research efforts will need to focus more on the development of technologies that will improve product quality and assure food safety. Recent advances in biotechnology such as the use of recombinant DNA technology to supply large quantities of polypeptide hormones such as somatotropin (ST) appear to hold considerable promise towards satisfying these changing demands for animal food products.

To ensure eventual consumer acceptance of these new technologies it will be critical to demonstrate that the technologies will: 1 – be cost effective, 2 – enhance nutritional value, 3 – improve product quality and 4 – be safe for human consumption, the target animals and the environment. Somatotropin is a polypeptide hormone secreted from the anterior pituitary that exerts control over numerous diverse metabolic processes. Although the exact modes of action for ST have yet to be fully elucidated, treatment of growing animals with ST has been shown to increase muscle growth and reduce adipose tissue stores. Somatotropin's actions on muscle growth are mediated indirectly via IGF-I to increase cellular proliferation and to increase whole-animal protein synthesis. The anabolic actions of ST on muscle and bone tissue, combined with the effects of ST on metabolism to redirect nutrient utilization, results in leaner tissue growth in meat animals.

RESEARCH DEVELOPMENTS LEADING TO USE OF RECOMBINANT ST

While the technology used to produce recombinant ST is relatively new, the research which led to the idea of using ST as an anabolic agent to enhance animal productivity is not. Studies conducted in the 1930s using crude preparations of pituitary ST in rats were the first to provide evidence to demonstrate the positive effect of ST on growth.

Recombinant DNA is a term which describes the splicing together or recombining of two or more pieces of DNA. The recombinant DNA techniques used to produce recombinant BST involves: the cloning of DNA (cDNA) from BST mRNA, the recombining of the BST cDNA with bacterial plasmid DNA and re-insertion of the "hybrid" DNA into *Escherichia Coli* bacteria for eventual expression of recombinant BST. The recombinant bacteria are grown in large numbers in typical fermentation vats, and the recombinant BST product extracted and purified. These same procedures are currently being used to produce human somatotropin and insulin for the treatment of dwarfism and diabetes mellitus, respectively.

Of the four recombinant BST products that have been produced by U.S. pharmaceutical companies and submitted for FDA approval for use in lactating cows, one is identical in amino acid sequence to endogenous BST, while the other three differ only slightly at the NH_2-terminus end of the protein. Recombinant BST is immunologically indistinguishable from endogenous BST using standard radioimmunoassay procedures.

The first experiment to be conducted with recombinant BST was a short-term study with lactating dairy cows which demonstrated that endogenous BST and recombinant BST were similar in increasing milk yields. The first experimental application of the use of recombinant ST to enhance growth was conducted with pigs.

COST-EFFECTIVENESS OF RECOMBINANT BST

Assuming recombinant BST were to be approved, the eventual adoption of recombinant BST by the beef industry and its cost effectiveness to the consumer will, among other things, be dependent upon: 1 – recombinant BST's ability to enhance growth performance and feed efficiency, 2 – the ability of recombinant BST to enhance lean tissue growth, 3 – consumer acceptance of leaner beef, 4 – market value of leaner beef and 5 – cost and delivery of recombinant BST.

The efficacy of recombinant BST for use as an anabolic agent in cattle to enhance the production of lean beef can not be fully accessed due to the limited amount of research that has been conducted with beef cattle to date.

An excellent 1989 review of the studies conducted through 1988 on the effects of BST on productivity in ruminants. Results of nine longer term trials (≥ 100 doses of BST treatment) concluded with cattle, indicate that BST treatment increased growth rates by approximately 12 percent with the range of responses being from 3 percent to 25 percent

EFFECT OF RECOMBINANT BST ON THE NUTRITIONAL VALUE OF BEEF PRODUCTS

Although the growth responses in cattle to recombinant BST have been variable, most studies have shown that BST is effective at increasing the lean:fat ratio in the carcass. In a recent study recombinant BST provided for 112 days increased daily protein gain by 16 percent and daily fat gain also tended to be lower (20 percent) in BST-treated steers. In other recent work, BST reduced the percent fat of the 9-10-11 rib section (indicator cut of carcass composition) by 13.4, 21.2 and 54.4 percent with increasing dosage of BST. Recombinant BST in our studies has also been shown to be effective in growing beef heifers, where 9-10-11 rib sections from BST treated heifers contained more protein (16.3 vs 18.3 percent) and less fat (26.0 vs 16.5 percent). Earlier research indicated that carcass leanness was increased by both BST and estradiol and that the responses were additive suggesting that the mode of actions for these anabolic agents may be independent. Recent research also suggests that the anabolic response to BST is very much dependent upon dietary protein and energy levels.

EFFECT OF RECOMBINANT BST ON THE SAFETY OF BEEF PRODUCTS

Since 1985, pharmaceutical companies seeking approval for the use of recombinant BST in lactating dairy cows have had FDA authorization to market meat and milk from BST-treated cows. This decision was based on their evaluation of extensive toxicological data and review of scientific literature concerning BST. As a result of consumer concern and due to general misunderstanding of human food safety of BST, FDA in an unprecedented action, released results of experimental data that had been submitted to the FDA by the pharmaceutical companies seeking recombinant

BST approval. The results of these studies were recently published and provide an excellent review of pertinent scientific studies which were used as the basis for the decision to authorize zero day withdrawal for the consumption of milk and meat from recombinant BST-treated cows.

For years it has been known that pituitary BST is ineffective when administered to humans. Although pituitary derived human ST and BST both have 191 amino acids, there is approximately a 35 percent difference in their amino acid sequence. The amino acid sequence of recombinant BST products and pituitary BST are either identical or differ by only a few amino acids at the NH_2 terminus. Because BST is a protein, it will be degraded in the gastrointestinal tract just like any other protein. The product of protein digestion (amino acids) generally enter the blood circulation almost entirely as free amino acids. In order to obtain authorization for a zero day withdrawal for the marketing of milk and meat from cows treated with BST, each pharmaceutical company seeking recombinant FDA approval had to conduct studies demonstrating the oral inactivity of their individual recombinant BST products.

INSULIN-LIKE GROWTH FACTOR I (IGF-I)

Another concern that has been raised with regards to potential harmful effects due to ingestion of BST-treated milk or meat involves IGF-I. Indeed, IGF-I levels are regulated by ST and elevated levels of IGF-I have been found in plasma, milk and meat of bST treated cows, and there exists 100 percent homology between bovine IGF-I and human IGF-I (identical amino acid sequences). However, these concerns have no scientific basis. First IGF-I, like ST and all proteins, will be digested to amino acids. Secondly, IGF-I levels in human breast milk have been shown to be seven to eight ng per ml and 19 ng per ml at six to eight week postpartum. These levels of IGF-I are much higher than IGF-I levels that have been found in milk obtained from cows treated with recombinant BST. Although pasteurization of milk will not destroy IGF-I levels, the heating process that milk is subjected to in the preparation of infant formula denatures all but approximately one tenth of the IGF-I present.

Excerpted From:

ANIMAL GROWTH BIOTECHNOLOGY: PST PERSPECTIVES-WHO BENEFITS?

Catherine M. Lemieux, Federal Reserve Bank, Kansas City

Advances in the biotechnology area are leading to the creation of products that will have a significant impact on agricultural production. For the pork sector, the Office of Technology Assessment (OTA) predicts that new technology will increase feed efficiency 12.6 percent and the number of pigs per sow by 27.6 percent between 1982 and 2000. This assumes an 80 percent adoption rate for new technologies.

Porcine somatotropin, PST, is one example of a biotechnology that is currently under development. It is a naturally occurring growth hormone in pigs that increases the rate of gain and feed efficiency of finishing hogs. Although it was discovered in the 1950s, it was not until the arrival of biotechnology that the protein could be manufactured in commercial quantities.

BACKGROUND

Somatotropin is a naturally occurring protein, produced in the pituitary gland of animals and humans, that regulates growth. Each species has its own unique somatotropin which is ineffective in other species. PST, for example, is inactive in humans, monkeys, and birds. Since somatotropins are proteins they are readily digestible and orally inactive. These findings are a plus for food safety because they mean that somatotropins do not pose a health risk in the human food chain.

What does PST do? Supplemental PST alters a pig's metabolism so that it converts feed energy to muscle more efficiently. This means that each pound of gain requires less feed and the resulting carcass contains more muscle and less fat.

Meisinger, in a summary of 19 studies, reported improvements in feed efficiency (unit of feed per unit of gain) of 24 percent, increases in rate of gain (units of gain per day) of 15 percent, and reductions in backfat of 25 percent for hogs treated daily with PST.

The implications of these studies are that leaner, heavier hogs can be produced using less feed in the same amount of time. Such an innovation has implications for consumers, packers, hog producers, and feed grain producers.

Porcine somatotropin by itself is size of operation neutral. Differences in response are more likely to be attributable to differences in management ability than size of operation. The benefits have been demonstrated on hogs with diverse genetic make-ups, but better quality animals show greater benefits. Improved nutrition also increases the impact of PST.

In spite of these apparent benefits the public is concerned about the use of somatotropins in general. There are important differences between the economic benefits of PST and BST. Use of PST allows farmers to produce a product that consumers are demanding, leaner meat. Efficiencies in production will be translated into lower consumer prices for leaner pork. BST increases feed efficiency in dairy cows and increases milk production. However, the product produced is not distinguishable from milk from any other cow. The only benefit to BST adoption is that the efficiency of milk production increases. Therefore, the economic implications of the adoption of PST are not readily transferable to the adoption of BST and vice versa.

Improvements in agricultural production have traditionally resulted in lower prices for farmers because costs of production have fallen and there has been little change in demand. Most advances in agricultural production have improved efficiency with no change in the final consumer product. PST is different. It not only increases the efficiency of production but also produces a more desirable consumer product and reduces the cost of processing that product by reducing the amount of fat that must be trimmed.

What are the benefits to consumers? The scientific evidence is that PST-treated hogs produce leaner carcasses which means that the retail cuts from treated hogs have less fat (both trim and intramuscular fat) and have fewer calories than untreated hogs. Consumer perceptions play a roll in demand. Consumers are concerned about food safety and residues as well as fat and calories. The fact that the National Institutes of Health have declared BST milk to be safe has not stopped consumer concern about BST milk. It is not known how consumers will accept PST.

Two U.S. consumer studies, one done by a market research firm in 1986 and the other done in 1991 at Iowa State University, and a 1989 British study have been analyzed. They yield surprisingly similar results. They show that consumers, given information on the safety of PST, are willing to pay a premium for the leaner cuts of pork from PST-treated hogs. A survey of over 200 households in the U.K. suggests that the British are generally willing to pay more (as a percentage of retail price) for leaner pork than U.S. consumers. This survey found that urban consumers, smaller families, and consumers at the bottom and top of the income range generally will pay higher premiums for leaner pork.

What benefits to packers transmit to producers? The large packers are currently paying premiums for leaner hogs. Cutability tests done by Prusa and Christian found that heavier, leaner carcasses from PST treated hogs yielded an extra $45.84 worth of retail cuts compared to control carcasses slaughtered at normal weights. Unless packers transmit consumers' increased demand for leaner pork and some of the cost savings in processing leaner pork back to the producer, a large part of the economic incentive for adoption of this new technology will be lost.

Based on the 1986 estimates of consumer demand for leaner meat, packers would be willing to pay 6.5 to 7.5 percent more for leaner carcasses. If hogs are $50.00 per hundredweight then packers would pay an additional $3.25 to $3.75 per hundredweight for leaner PST hogs.

These estimates represent the portion of the premium consumers are willing to pay for leaner meat that will be shared with producers. This sharing is not a result of the benevolence of the packing industry but a direct result of their quest for market share and profits.

How will PST impact the hog industry? Lemieux and Wohlgenant developed a model of the hog industry that estimates the impact of the introduction of PST on farm price of hogs, farm quantity of hogs produced, hogs slaughtered, retail price of pork, retail consumption of pork, hog imports, hog exports, economic benefits to producers and the economic benefits to consumers. Based on these relationships the production benefits of PST must be translated into shifts in industry supply.

For this study PST was assumed to increase feed efficiency 24 percent (from a base of 3.87 feed per gain) and the rate of gain increased 15 percent (from a base of 1.7 gain per day).

The improvement in rate of gain means that the control pig gained 100 pounds and was marketed at 240 pounds, while the PST pig gained 115 pounds and was marketed at 255 pounds (assuming a constant 59 day finishing period). This additional gain amounts to an increase in production of 6.3 percent.

Costs of PST treatment must be deducted from the gains. In this study PST cost was assumed to be $6.00 per animal. With a market hog selling for $52 per hundredweight, total treatment costs are five percent of the value of the hog.

These results indicate that a typical producer could expect to increase per unit profit between 10.1 percent (19 percent cost decrease less 8.9 percent price decrease) and 11.9 percent (19 percent cost decrease less 7.1 percent price decrease) if 60 percent of the industry adopts PST. This compares with decreases in per unit profits of between 7.1 and 8.9 percent for nonadopters.

The original question was, "How will PST impact the hog industry?" Producers' surplus calculates the net increase in profits for the industry aggregated over adopters and nonadopters and is estimated to increase between 2.5 to 4.4 percent of the total value of hog production. In 1987, the total value of hog production was about $10 billion indicating increases in the benefits to hog producers would range between $250 million to $440 million after five years. The net benefits to consumers ranged between 9 and 11.5percent of the total value of hog production. The dollar value of these benefits using the 1987 value of hog production is between $900 million and $1.15 billion.

These benefit levels are sensitive to adoption rates by producers, length of time of adjustment, and whether demand for pork increases in response to introduction of PST.

What impact will PST have on individual producers? To investigate this question, three typical farms were developed for each type of operation (farrow-to-finish and hog only) from information obtained from USDA. These farms feed approximately 700, 1700 and 3,400 hogs per year and reflected average efficiencies within their respective size group.

All farms simulated received at least a 150 percent return on investment. Benefits were larger for producers that had to purchase all their feed than for producers that grew their own feed because the additional income received through increased crop sales on the farrow-to-finish farms was less than the value of feed saved on the hog only farms. The economic benefits of producing leaner carcasses provided the majority of the economic benefits.

CONCLUSIONS

Benefits this technology provides for the economy as a whole and to individual segments are significant. No real losers can be identified.

Consumers will be able to purchase safe, less expensive, lean pork. When the impact of increased supply and leaner retail products is quantified, the estimated benefits range between $900 million and $1.15 billion. Producers will be marketing a product that is in greater demand and will cost less to produce.

Packers will be able to reduce processing costs by buying leaner hogs. Each PST hog will provide greater higher priced retail cuts, such as roasts and hams, increasing the return to the packer for each animal processed. Research on the profitability of PST adoption show that a majority of the benefits of adoption come from this leanness premium.

Producers that adopt PST can produce more with less. Less feed will be needed to produce heavier hogs in the same amount of time that it currently takes.

More than ever, the public seems to mistrust research on the safety of new technologies. Consumer studies show that the identification of a product as a "hormone" causes a significant negative bias. The estimates assume consumers will not respond negatively to PST pork. Unless consumers are educated about the safety of the product and its potential benefits, public concerns could limit the adoption of this technology.

SUMMARY

PST is a naturally occurring protein that stimulates growth in pigs but is inactive in humans. It has been found to increase rate of growth, improve feed efficiency, and increase leanness in finishing hogs. With PST, consumers will be able to purchase leaner pork at lower prices. When the value of these consumer benefits is calculated the estimates range between $900 million to $1.5 billion. Producers will see industry profits increase $250 million to $440 million. These benefits will be available to all size producers and profit per pig does not increase with the size of the operation. Use of PST increases profits for all size operations, types of farms, regardless of the farm programs investigated.

Excerpted in part from:

BETA AGONISTS

David. P. Hutcheson, Agricultural Research and Extension, Texas A&M University System

"Beta agonists" refers to a family of compounds that include synthetic beta-adrenergic agonists which are similar to catecholamines. Beta- agonists, thus, are involved in different metabolic actions. The beta-agonists of interest are compounds that decrease anabolism of lipids and allow greater anabolism of protein. These types of beta agonists are in line with present production goals of producing more lean and less fat.

At the present time, the beta agonists that are of interest are ractopamine, salbutamol, clenbuterol, cimaterol and L-6449,969. Activity of these compounds covers several species, including beef cattle, swine and sheep. While none of these have yet been approved for use commercially, administration has been through the feed in studies conducted.

Early studies indicated clenbuterol was effective in reducing fat and increasing lean in cattle and sheep, but it is no longer being considered as a potential growth regulator.

Recent studies with cimaterol and ractopamine indicate potential for swine with some responses in cattle and sheep. Studies with swine indicate ractopamine and cimaterol are effective means of reducing fat deposition and enhancing lean pork production.

One of the concerns about using products like beta agonists is that they will dramatically increase the quantity of feed, especially protein, required by swine. This cost increase could conceivably offset any benefit from using the products. However, current research has indicated that elevated dietary protein was not necessary to achieve an increase in carcass leanness. Other research has found an increase in efficiency of feed utilization by swine receiving ractopamine, whereas, cimaterol had no effect on feed utilization. Together, this information demonstrates that, at least with ractopamine, the increase in carcass leanness will not require additional dietary protein, and may reduce the total feed requirements to produce a unit of meat by finishing swine.

A summary of six beef cattle trials with ractopamine fed the last 38 to 46 days before slaughter, indicated improved average gain (18.4 percent) and decreased feed to gain (15.7 percent) with a slight reduction of 4.9 percent carcass percent fat and an increase of 2.5 percent carcass percent protein.

Beta agonists allow the restriction of fat during the growing and finishing period of animals. With less production of fat, a more desirable protein product can be produced. The producer of should benefit the most from beta agonists since these compounds should reduce the cost of production.

The consumer, whether at home or at a restaurant, determines what technologies are ultimately used in production of lean red meat. Presently, the consumer prefers lean red meat to higher levels of fat in the meat, but the consumer has equated tenderness to fat levels in red meat. Technologies such as beta agonists offer the potential to produce a quality protein product that meets consumer demands at a lower production cost.

PARTICIPANTS

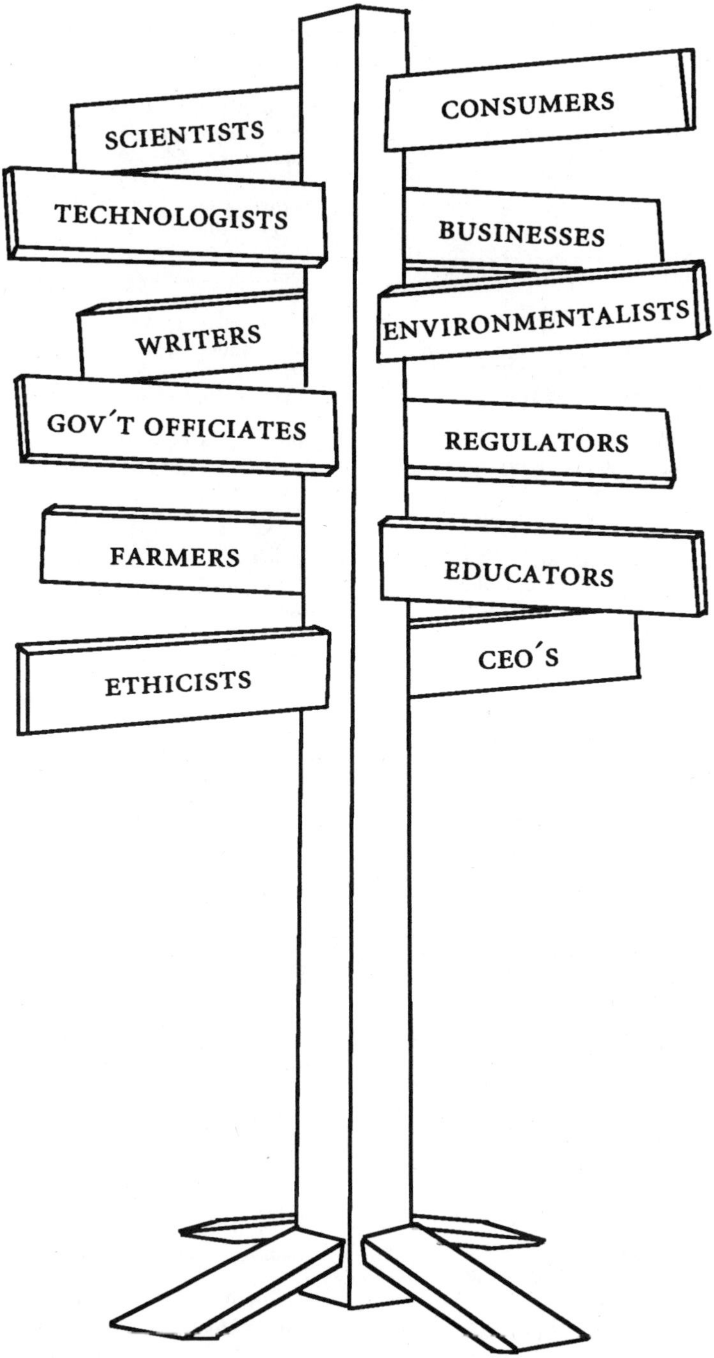

LIST OF PARTICIPANTS

James J. Ahle
ICI Americas, Inc.
1200 South 47th Street
Richmond, CA 94804

Chris Aggour
Center for Public Issues in Biotechnology
Univ. of Maryland, 2206 Symons Hall
College Park, MD 20742–5535

Larry J. Allen, DVM
CA Dept of Food and Agriculture
1220 N Street, Room A 107
Sacramento, CA 95814

John Algeo
Cal Poly State University
1 Grand Avenue
San Luis Obispo, CA 93407

Larry J. Allen, DVM
CA Dept of Food & Agriculture
1220 N Street, Room A–107
Sacramento, CA 95814

Jim Altemus
Monsanto
800 North Lindbergh
St. Louis, MO

Walter Truett Anderson
1112 Curtis Street
Albany, CA 94706

Joe Antognini
USDA ARS NPS, Rm 237, Bldg 005
BARC West
Beltsville, MD 20705

Charles J. Arntzen
Deputy Chancellor for Agriculture
The Texas A&M University System
College Station, TX 77843–2142

Elizabeth Atcheson, Vice President
Center for Science Information
4252 20th Street
San Francisco, CA 94114

Brian Baker
CA Certified Organic Farms
P.O. Box 8136
Santa Cruz, CA 95061

Kenneth A. Barton
Agracetus, Inc.
8520 University Green
Middleton, WI 53562

Fernando Bas
Catholic University of Chile
Casilla 6177
Santiago, 22 Chile

James J. BatesICI Americas
1200 South 47th Street
Box 4023
Richmond, CA 94804

Bill R. Baumgardt
Director, Ag Exp Sta.,Purdue Univ.
116 AGAD
West Lafayette, IN 47907

Alan B. Bennett
Vegetable Crops, Univ of CA
Davis, CA 95616

Richard Bennett
Dairy & Public Policy Advisor
2604 Ventura Avenue, Rm 100–P
Santa Rosa, CA 95403–2894

Stephen Benoit
Calgene, Inc.
1920 Fifth Street
Davis, CA 95616

Karen Bernstein
San Mateo, CA 94401

Edgar Beyer
Cal Poly State University
San Luis Obispo, CA 93407

James Brenner
4–H Youth & Human Resource ADvisor
2604 Ventura Avenue, Rm 100
Santa Rosa, CA 95403

Winston J. Brill
Winston J. Brill & Associates
4134 Cherokee Drive
Madison, WI 53711

Edward Bruggemann, Staff Scientist
National Audubon Society
801 Pennsylvania Ave., SE
Washington, DC 20003

Clark R. Burbee
CSRS USDA
STE 329, Aerospace Center
Washington, DC 20250–2200

L. J. "Bees" Butler
Agricultural Economics, Univ of CA
Davis, CA 95616

Floyd M. Byers
Animal Science
230 Kleberg Hall, Texas A&M Univ.
College Station, TX 77843

John Callahan
Calgene, Inc.
1920 Fifth Street
Davis, CA 95616

Dwight Camper
Plant Pathology & Physiology
Clemson University
Clemson, SC 29634–0377

Harold O. Carter, Director
UC Ag Issues Center, UC Davis
Davis, CA 95616

Jerry D. Caulder,President &
CEOMycogen Corporation
5451 Oberlin Drive
San Diego, CA 92121

Brian F. Chabot
Director of Research
Ag Experiment Sta., Cornell University
Ithaca, NY 14853

Judith Chambers
US AID, Office of Agriculture
Room 406–A, SA–18
Washington, DC 20523

Neville P. Clarke, Director at Large
SES, Texas A&M University
Room 308, Systems Admin. Building
College Station, TX 77843–2129

Glenn B. Collins
University of Kentucky
N 222 K Agri. Sci. N; Agronomy
Lexington, KY 40546–0091

Jane Harris Cramer
Agrigenetics Company
5649 East Buckeye Road
Madison, WI 53716

Lisa Crumrine
Fresno Bee
1626 East Street
Fresno, CA 93786

Elwood Dart, N. Div. Sales Manager
Wilbur–Ellis Company
P.O. Box 896
Moses Lake, WA 98837

Peter R. Day, Director
Center for Ag. Molecular Biology
Rutgers; Cook College
New Brunswick, NJ 08903

Edward Debus
Monsanto Agricultural Company
700 Chesterfield Village Parkway
St. Louis, MO 63198

Joseph W. Deverna Manager, Veg.Biotech
Campbell Soup Company
28605 Road 104
Davis, CA 95616

Charles Divan
USDA BBEP, Room 828
6505 Belcrest Road
Hyattsville, MD 20782

R. N. Dryden Jr.
Big Stone Inc.
885 Arapahoe Avenue
Boulder, CO 80302–6029
Stanislaus J. Dundon
626 B Street
Davis, CA 95616

Peter E. Dunn, Director
Ag Biotech, Purdue University
1158 Entomology Hall
West Lafayette, IN 47907–1158

Dennis Edwards
Entotech
1497 Drew Avenue
Davis, CA 95616

Matt Elhardt
Hoeschst–Roussel Agri–Vet
268 Via Mission Drive
Chico, CA 95928

Don Emlay
Calgene, Inc.
1920 Fifth Street
Davis, CA 95616

Ben Faber
Farm Advisor, UC Coop Ext
800 S. Victoria Avenue, L128
Ventura, CA 93009

Jeffrey C. Fearn, NABC/Joyce Fellow
Boyce Thompson Inst, Plant Prod Res.
Tower Road
Ithaca, NY 14853–1801

Gail Feenstra
UC Sustainable Ag. Res. & Ed. Program
UC Davix, 258 Hunt Hall
Davis, CA 95616
Walter R. Fehr
Director of Biotechnology
I SU, 1010 Agronomy
Ames, IA 50011

June Fessenden MacDonald
Deputy Director, NABC
159 Biotechnology Building
Cornell University
Ithaca, NY 14853–2703

D'Ann Finley, Coop Ext.
UC Davis
Davis, CA 95616
Neal First
Madison, WI

Douglas Fisher
DeKalb Genetics Corporation
3100 Sycamore Rd.
DeKalb, IL 60115

Eric Flamm, Senior Staff
FDA, Office of Biotechnology
5600 Fishers Lane
Rockville, MD 20857

Joseph R. Fordham, Director
Reg. Affairs, Novo Nordisk Bioindustrials
33 Turner Road
Danbury, CT 06813–1907

Barney Gallagher
Sandoz Crop Protection Corporation
975 California Avenue
Palo Alto, CA 94304

Robert G. Gast, Director
Ag Exp Stat, Michigan State University
109 Agriculture Hall
East Lansing, MI 48824–1039

Rebecca Goldburg, Biologist
Environmental Defense Fund
275 Park Avenue, South
New York, NY 10010

Ben Gonzales
Office of Health & Safety
UC Berkeley, University Hall
Berkeley, CA 94720

Nelson Goodman
USDA ARS, 800 Buchanan Street
Albany, CA 94710

Robert R. Granados, Program Director
Plant Protection, Boyce Thompson Inst.
Tower Road
Ithaca, NY 14853–1801

Ron Gregory
Dow Elanco, P.O. Box 129
Penryn, CA 95663

John Griesbach
Plant Pathology, Oregon Dept. of Ag.
635 Capitol NE
Salem, OR 97310–0110

James Grieshop
Applied Behaviral Research, UC Davis
Davis, CA 95616

Harold Hafs, Vice President
An Health & Ag
Merck Sharp & Dohme Res. Labs
P.O. Box 2000 WBC 120
Rahway, NJ 07065

Michael K. Hansen
Project Leader, Consumers Union
Research Associate, Consumer Policy Inst.
Institute for Consumer Policy Research
101 Truman Avenue
Yonkers, NY 10703–1087

Ralph W. F. Hardy, Chair, NABC
Pres., Boyce Thompson Instfor Plant Res
Tower Road
Ithaca, NY 14853–1801

Charles E. Hess, Asst. Sec. for Sci & Ed
USDA Room 217–W, Admin Bldg
Washington, DC 20250

Robert A. Hill
Doane Marketing Research
1807 Park 270 Drive, Suite 300
St. Louis, MO 63146

Maureen Hinkle, Director of Ag Policy
National Audubon Society
801 Pennsylvania Avenue, SE
Washington, DC 20003

Catherine Houck
Calgene, Inc.
1920 Fifth Street
Davis, CA 95616

Theodore L. Hullar, Chancellor
UC Davis
Davis, CA 95616

Susanne L. Huttner
UC Systemwide Biotech Program
405 Hilgard Ave., MBI Bldg. Room 103
Los Angeles, CA 90024–1570

James Hyer
#116 An Sci Dept, Caly Poly State Univ
San Louis Obispo, CA 93406

Ernestine Ivans
UC Cooperative Extension
680 North Campus
Hanford, CA 93323

Dixie Jacobsen
Calgene Inc.
1920 Fifth Street
Davis, CA 94515

Alvin Janski
Pitman–Moore, Inc.
Box 207
Terre Haute, IN 47808

Alfred L. Jin, Office of the President
UC Oakland, 300 Lakeside Drive, 7th Fl.
Oakland, CA 94612–3550

Bryan Johnson, Associate Director
Ag ExpSta, Univ. of Maryland
Room 1118, Symons Hall
College Park, MD 20742

Charles A. Johnson
Texas A&M University
Harrington Building, Room 801
College Station, TX 77843

Nandini Katre
Katre Consulting
6107 Jordan Avenue
El Cerrito, CA 94530

Lynne Kennedy, Conference Coordinator
Ag Issues Center, UC Davis
Davis, CA 95616

John E. Kinsella
Ag & Env Sci., U C Davis
Davis, CA 95616

Cassandra Klotz, NABC/Joyce Fellow
UC Davis, Agricultural Economics
Davis, CA 95616

Brewster Kneen
125 Highfield Road
Toronto, Ontario M4L 2V4 CANADA

Don Koehler
CA Dept. of Food and Agriculture
1220 N Street., Rom A–400
Sacramento, CA 95814

Dave Krantz
CA Farm Network & CA Farm Bureau
1601 Exposition Blvd., FB10
Sacramento, CA 95815

Joan Kureczka
Bio Venture View (press)
380 Pennsylvania Avenue
San Francisco, CA 94107

William B. Lacy, Asst. Dean for Research
229 Ag Admin, PA State Univ
University Park, PA 16802

Alvin Lazen
Committee. on Life Sciences
National Academy of Sciences
2101 Constitution Ave., NW
Washington, DC 20418

Peggy G. Lemaux
Plant Biology, UC Berkeley
111 Genetics & Plant Biology Building
Berkeley, CA 94720

Bill Leicht
P.O. Box 10094
Fresno, CA 93745

William B. Lesser
Ag Ec, 310 Warren Hall
Cornell University
Ithaca, NY 14853

Les Levidow
The Open University
48 Fortress Road
London NW 52HG ENGLAND

Randolph V. Lewis
University of Wyoming
P.O. Box 3944
Laramie, WY 82071–3944

Ellen Liberman
USDA BBEP
6505 Belcrest Road
Hyattsville, MD 20782

Bill Liebhardt
Sustainable Ag Program, UC Davis
Davis, CA 95616

Jim Lyons
Div of Ag & Nat. Resources, UC Davis
IPM Project
Davis, CA 95616

David R. MacKenzie, Director
Nat Biological Impact AssessmentProg.
USDA
901 D Street, SW, Suite 330
Washington, DC 20250–2200

Lori Malyj
Calgene, Inc.
1920 Fifth Street
Davis, CA 95616

Mark Mammarella
OEHS, UC Berkeley, University Hall
Berkeley, CA 94720

Pamela Marrone, President
Entotech
1497 Drew Avenue
Davis, CA 95616

Peter Y. Matlock
P. Matlock Biotechnology Development
1306 Rosario Street
Davis, CA 95616

Jenifer Huang McBeath
Ag & Forestry Exp Sta, Univ of Alaska
Fairbanks, AK 99775–0080

Carol McClain, Office of the President
UC Oakland, 300 Lakeside Drive; 18th Fl
Oakland, CA 94612–3550

Martina McGloughlin
Biochemistry & Biophysics, UC Davis
Davis, CA 95616

Murray McLaughlin
Ag–West Biotech Inc.
105–15 Innovation Blvd.
Saskatoon, Sask. S7N 2X8 CANADA

Valerie Mellano
UC Coop Ext., San Diego County
5555 Overland Avenue, Building #4
San Diego, CA 92123

Margaret G. Mellon, Director, National
Biotechnology
Policy Center
National Wildlife Federation
1412 Sixteenth Street, NW
Washington, DC 20036

Ken Mishra
Director, Plant Breeding Division
Pioneer Hi–Bred International, Inc.
6800 Pioneer Parkway
P.O. Box 316
Johnston, IA 50131

Jim Morris
California Farm Network
California Farm Bureau
1601 Exposition Blvd. FB10
Sacramento, CA 95815

Stephen Muench, Manager, Product Dev.
Monsanto Company
800 North Lindbergh Boulevard
St. Louis, MO 63167

Leslie Murchy, PA Consultant
950 Steiner Street, #3
San Francisco, CA 94117

George F. Murphy
Taraval Associates
1220 University Drive # 204
Menlo Park, CA 94025

James D. Murray
Animal Science, UC Davis
Davis CA 95616

K. Darwin Murrell, Director
Midwest Area, ARS/USDA
1815 University
Peoria, IL 61604

Carolyn Napoli
Environmental Horticulture, UC Davis
Davis, CA 95616

Gordon Neish
Ag Canada Research Station
P.O. Box 3000 Main
Lethbridge, Alberta T1J 4B1 CANADA

Aaron Nelson
UC Cooperative Extension
1720 South Maple Avenue
Fresno, CA 993702

Merritt R. Nelson
Plant Pathology, Univ of Arizona
Tucson, AZ 85721

William E. Newton
Virginia Tech Ag Biotech Center
123 Engel Hall
Blacksburg, VA 24061–0308
Robert B. Nicholas, Exec Director, NABC
Partner, McDermott, Will & Emery
1850 K Street, NW, Suite 500
Washington, DC 20006–2296

Susan E. Offutt
Chief Agriculture Branch, OMB
New Executive Office Building,Rm 8025
Washington, DC 20503

Brian Orshinsky
Saskatchewan Wheat Pool
c/o NRC—PBI
Saskatoon, Sask. S7K 6W9 CANADA

Nancy Ostiguy
CSUS, 1412 24th Street
Sacramento, CA 95816

Douglas D. Parker
University of Maryland
CPIB, 2206 Symons Hall
College Park, MD 20742

Patrico Parodi
Catholic University of Chile
Casilla 306
Santiago, 22 CHILE

Gary Patton, County Supr.
Third (Seaside) District
701 Ocean Street
Santa Cruz, CA 95060

William Pengelly
N SF, 1800 G Street, NW
Washington, DC 20550

Tarla Peterson
Texas A&M Univ,Speech Commun.
College Station, TX 77845

Louis J. Pierro, Associate Director
Storrs Ag Exp Sta, Univ of CT, U–10
1376 Storrs Road
Storrs, CT 06829–4010

Roley Piggott, NABC/Joyce Fellow
Agricultural Economics, UC Davis
Davis, CA 95616

Bill Plummer
Cal Poly State University
San Luis Obispo, CA 93407

Cynthia Puffer
Resources for the Future
1616 P Street, NW
Washington, DC 20036

Calvin O. QualsetDirector
Genetic ResConservPro, UC Davis
Davis, CA 95616

Jo Randall
ICI/Garst Seed Company
POB 500
Slater, IA 50244

Keith Redenbaugh
Calgene, Inc.
1920 Fifth Street
Davis, CA 95616

James Reedy, Comm Res Adv.
UC Coop Ext
2000 West Texas Street
Fairfield, CA 94533–4498

W. R. Reinert, Director
Veg Research, Campbell Soup Company
28605 Road 104
Davis, CA 95616

Caird E. Rexroad Jr.
USDA ARS BARC East,Bldg 200, Rm 13
10300 Baltimore Avenue
Beltsville, MD 20705

Ellen Rilla, County Director
UC Coop Extension
1450 A Lucas Valley Road
San Rafael, CA 94903

Benito Alvarado Rodriguez
Apdo Postal #185
Guasave, Sinaloa, MEXICO 81000

Susan Rossi
Sogetal, Inc.
3876 Bay Center Place
Hayward, CA 94545
Roger H. Salquist, Chairman & CEO
Calgene, Incorporated
1920 Fifth Street
Davis, CA 95616

Ruedi Sandmeier, V P, Crop Research
Sandoz Crop Protection
975 California Avenue
Palo Alto, C 94304–1104

Henry Schacht
60 Hiller Drive (press)
Oakland, CA 94618

Rhonda Scott
Amoco Technology company
150 West Warrenville Road, MS–E6
Naperville, IL 60566

George E. Seidel Jr.
Animal Physiology, CSU
Fort Collins, CO 80523

John A. Shadduck
Ass. Deputy Chancellor & Dean
College of Veterinary Medicine
The Texas A&M University System
College Station, TX 77843–4461

Sano Shimoda
Montgomery Securities
600 Montgomery street
San Francisco, CA 94523

Oystein Simonsen
Agricultural Research Council of Norway
1005 F Street, P.O. Box 367
Davis, CA 95616

Ann Sorensen
American Farm Bureau Federation
225 Touhy Avenue
Park Ridge, IL 60068

Doreen Stabinsky
Plant Pathology, UC Davis
Davis, CA 95616

Gerald G. Still
USDA ARS Plant Gene Expression Center
800 Buchanan Street
Albany, CA 94710

Ralph Stoaks
USDA APHIS PPQ
9580 Micron Avenue, Suite I
Sacramento, CA 95827

James W. Stratton
CA Dept. of Health Services
714 P Street, Room 450
Sacramento, CA 95814

Paul K. Stumpf
USDA CSRS NRICGP
323 Aerospace Bldg.
Washington, DC 20250–2200

Patricia B. Swan, Vice Provost
Research & Advanced Studies, ISU
211 Beardshear
Ames, IA 50011

Johan Swinnen, NABC/Joyce Fellow
Ag Ec, Cornell University
Ithaca, NY 14853

Martin Terry
Animal Health Institute
119 Oronoco Street
Alexandria, VA 22314–2058

Paul B. Thompson, Director
Center for Biotechnology Policy & Ethics
Texas A&M University
014 Bolton
College Station, TX 77843–4355

Linda Tombras Smith
University of California
228 Mrak Hall
Davis, CA 95616

Janet Toppenberg
California Farm Bureau Federation
1601 Exposition Blvd.
Sacramento, CA 95616

Ellen C. Truong
Humane Society of the U.S.
2100 L Street, NW
Washington, DC 20037

Seymour D. Van Gundy
Nematology, UC Riverside
3112 College Building, North
Riverside, CA 92521

Jane Voichick
University of Wisconsin
1415 Linden Drive
Madison, WI 53706

Carol Waddington
Tri- Cal Bio-Systems
P.O. Box 1327
Hollister, CA 95024

William H. Walker, III
Agricenter International
7777 Walnut Grove Road
Memphis, TN 38120

Ben Warner
Warner Public Relations (press)
2780 Marty Way
Sacramento, CA 95818

Malcolm Watts
Watts & Whelan Capital Management
456 Montgomery Street, Suite 900
San Francisco, CA 94104

David J. Webber
Political Science, Univ. of Missouri
316 Professional Building
Columbus, MO 65211

David Westenbarger
ERS, USDA
1301 New York Avenue, NW, Suite 532
Washington, DC 20005

Donald K. Weymouth, DVM
Lilly Research Laboratories
2001 West Main Street, POB 708
Greenfield, IN 46140

Stanley P. Wilson
Council for Ag. Science & Technology
137 Lynn Avenue
Ames, IA 50010–7197

Steven C. Witt, President
Center for Science Information
4252 20th Street
San Francisco, CA 94114

Zachary Wochok
Calgene, Inc.
1920 Fifth Street
Davis, CA 95616

Marcia Wood
USDA ARS
800 Buchanan Street
Albany, CA 94710

Melvin Woods
Manager, Public Affairs
Eli Lilly & Company
555 Capitol Mall, Suite 460
Sacramento, CA 95814

John Woolcott
Agricultural Issues Center, UC Davis
Davis, CA 95616

Roger Wrubel
Tufts University
97 Talbot Avenue
Medford, MA 02155

S. F. Yang
Mann Laboratory, UC Davis
Davis, CA 95616

Lawrence Yee
County Director
UC Cooperative Extension
800 S. Victoria Avenue, L128
Ventura, CA 93009

Curt Youngs
Animal Science, ISU
Ames, IA 50011

Milton Zaitlin
Plant Pathology
130 Biotechnology Bldg, Cornell Univ.
Ithaca, NY 14853–2703

Peg Zenk
Farm Industry News Magazine
7900 International Drive
Minneapolis, MN 55425

Lydia Zepeda
Ag Ec,University of Wisconsin
427 Larch Street
Madison, WI 53706